A Beginner's Guide to Mathematical Logic

Raymond M. Smullyan

Dover Publications, Inc.
Mineola, New York

Copyright

Copyright © 2014 by Raymond M. Smullyan
All rights reserved.

Bibliographical Note

A Beginner's Guide to Mathematical Logic is a new work, first published by Dover Publications, Inc., in 2014.

International Standard Book Number
ISBN-13: 978-0-486-49237-7
ISBN-10: 0-486-49237-0

Manufactured in the United States by LSC Communications
49237006 2017
www.doverpublications.com

Table of Contents

Part I
General Background

1
Genesis

Just what is Mathematical Logic? More generally, what is Logic, whether mathematical or not? According to Tweedledee in Lewis Carroll's *Through the Looking Glass*, "If it was so, it might be, and if it were so, it would be, but since it isn't, it ain't. That's logic."

In *The 13 Clocks* by James Thurber, the author says, "Since it is possible to touch a clock without stopping it, it follows that one can start a clock without touching it. That is logic, as I understand it."

A particularly delightful characterization of logic was given by Ambrose Bierce, in his book *The Devil's Dictionary*. This is really a wonderful book that I highly recommend, which contains such delightful definitions as that of an egotist: "An egotist is one who thinks more of himself than he does of me." His definition of logic is "Logic; n. The art of thinking and reasoning in strict accordance with the limitations and incapacities of the human misunderstanding. The basis of logic is the syllogism, consisting of a major and a minor premise and a conclusion thus:

Major Premise: Sixty men can do a piece of work sixty times as quickly as one man.
Minor Premise: One man can dig a post-hole in sixty seconds. Therefore,
Conclusion: Sixty men can dig a post hole in one second."

The philosopher and logician Bertrand Russell defines mathematical logic as "The subject in which nobody knows what one is talking about, nor whether what one is saying is true."

Many people have asked me what mathematical logic is, and what its purpose is. Unfortunately, no simple definition can give one the remotest idea of what the subject is all about. Only after going into the subject will its nature become apparent. As to *purpose*, there are many purposes, but again, one can understand them only after some study of the subject. However, there is one purpose that I can tell you right now, and that is to make precise the notion of a *proof*.

I like to illustrate the need for this as follows: Suppose a geometry student has handed in to his teacher a paper in which he was asked to give a proof of, say, the Pythagorean Theorem. The teachers hands back the paper with the comment, "This is no proof!" If the student is sophisticated, he could well say to the teacher, "How do you know that this is not a proof? You have never defined just what is meant by a *proof*! Yes, with admirable precision, you have defined geometrical notions such as triangles, congruence, perpendicularity, but never in the course did you define just what is meant by a *proof*. How would you *prove* that what I have handed you is not a proof?"

The student's point is well-taken! Just what is meant by the word *proof*? As I understand it, on the one hand it has a popular meaning, but on the other hand, it has a very precise meaning, but only relative to a so-called *formal* mathematical system, and thus the meaning of *proof* varies from one formal system to another. It seems to me that in the everyday popular sense, a proof is simply an argument that carries conviction. However, this notion is rather subjective, since different people are convinced by different arguments. I recall that someone once said to me, "I can *prove* that liberalism is an incorrect political philosophy!" I replied, "I'm sure you can prove this to your satisfaction, and to the satisfaction of those who share your values, but without even hearing your proof, I can assure you that your so-called *proof* would carry not the slightest conviction to those with a liberal philosophy!" He then gave me his "proof," and indeed it seemed perfectly valid to him, but obviously would not make the slightest dent on a liberal.

Speaking of logic, here is a little something for you to think about: I once saw a sign in a restaurant which read, "Good food is not cheap. Cheap food is not good."

Problem 1. Do those two statements say different things, or the same thing?

Note that solutions to problems are given at the end of the chapters.

Mathematical Logic is sometimes also referred to as *Symbolic Logic*. Indeed, one of the most prominent journals on the subject is entitled "The Journal of Symbolic Logic." How did the subject even start? Well, it was preceded by logic of a non-symbolic nature. The name *Aristotle* obviously comes to mind, for that famous ancient Greek philosopher was the person who introduced the notion of the syllogism. It is important to understand the difference between a syllogism being *valid* and a syllogism being *sound*. A valid syllogism is one in which the conclusion is a logical consequence of the premises, regardless of whether the premises are true or not. A *sound* syllogism is a syllogism which is not only valid, but in which, in addition, the premises are true. An example of a sound syllogism is the well-known:

All men are mortal.
Socrates was a man.

Therefore, Socrates was mortal.

The following is an example of a syllogism which, though obviously unsound, is nevertheless valid:

All bats can fly.
Socrates is a bat.

Therefore, Socrates can fly.

Clearly, the minor premise (the second one) is false, and also, of course, so is the conclusion. Nevertheless, the syllogism is valid – the conclusion really is a logical consequence of the two premises. If Socrates were a bat, then he really could fly.

I am amused by syllogisms that appear to be obviously invalid, but are actually valid! Here are two examples:

Everyone loves my baby.
My baby loves only me.

Therefore, I am my own baby.

Doesn't that sound ridiculous? But it really is valid! Here is why:

Since *everyone* loves my baby, then my baby, being also a person, loves my baby. Thus my baby loves my baby. But also, my baby loves only me (minor premise). Since my baby loves *only* me, then there is only one person my baby loves (namely me), but since my baby loves my baby, that one person must be me. Thus I must be my own baby.

Here is another example of a valid syllogism which seems invalid. We define a *lover* as one who loves at least one person.

Everyone loves a lover.
Romeo loves Juliet.

Therefore: Iago loves Othello.

Here is why the syllogism is valid. Since Romeo loves Juliet (second premise), then Romeo is a lover. Since Romeo is a lover, then everyone loves Romeo (by the first premise). Since everyone loves Romeo, then everyone is a lover. Since each person is a lover, then everyone loves that person (by the first premise). Thus it follows that everyone loves everyone! In particular, Iago loves Othello.

Once, the eminent logician and philosopher Bertrand Russell was asked, "What is really new in the conclusion of a syllogism?" Russell replied that *logically*, the conclusion may contain nothing new, but the conclusion can nevertheless have *psychological* novelty, and he then told the following story to illustrate his point.

At a certain party, one man told a somewhat risqué story. Someone else told him, "Please be careful. The abbot is here!" The abbot then said, "We men of the cloth are not as naïve as you might think! Why, the things I have heard in the confessional . . . my first penitent was a murderer!" Shortly after, an aristocrat came late to the party, and the host wanted to introduce him to the abbot, and asked if he knew him. The aristocrat said, "Of course I know him! I was his first penitent."

Aristotelian logic flourished through the ages. In the seventeenth century the philosopher Leibnitz envisioned the possibility of a symbolic calculating machine that would settle all questions – mathematical, philosophical and even sociological. Wars would then be unnecessary, since the opposing sides could instead say, "Let us sit down and calculate."

There is the story told that Leibnitz was undecided whether or not to marry a certain lady, and so he made one list of advantages and one of disadvantages. The disadvantage list turned out to be longer, and so he decided not to marry her.

As we will see in the course of this volume, an amazing discovery of the logician Kurt Gödel [1931] showed that Leibnitz' dream was unrealizable even in pure mathematics.

Symbolic Logic proper can be said to have its beginnings in the nineteenth century, through such thinkers as Pierce, Jevons, Schroeder, Venn, De Morgan and particularly George Boole, after whom Boolean Algebra is named. Boole was a completely self-educated school master, who wrote *An Investigation of the Laws of Thought* [republished, 2009], in which he described his purpose in writing the book in the following opening words:

> "The design of the following treatise is to investigate the fundamental laws of the mind by which reasoning is performed, to give expression to them in the symbolic language of a Calculus, and upon this foundation to establish the science of Logic and construct its method; to make that method itself the basis of a general method for the application of the mathematical doctrine of Probabilities, and finally, to collect from the various elements of truth brought to view in the course of these inquiries some probable intimations concerning the nature and constitution of the human mind."

The book is an interesting mixture of precise mathematical and symbolic reasoning with philosophical considerations. Boole attempts to put purely philosophical arguments into symbolic form – especially in his chapter on the philosophers Clarke and Spinoza. Towards the beginning of the chapter in which he does that, he says,

> "In the pursuit of these objects it will not devolve upon me to inquire, except accidentally, how far the metaphysical principles laid down in these celebrated productions are worthy of confidence, but only to ascertain what conclusions may justly be drawn from the given premises."

Thus Boole's purpose in that chapter is not to decide whether the philosopher's premises (and hence also the conclusions) are true, but only whether the conclusions are really logical consequences of the premises – in other words, not if the philosopher's arguments are *sound*, but only whether they are valid.

Boole becomes quite philosophical in the last chapter of the book. To my great delight he writes the following beautiful words:

"If the constitution of the material frame is mathematical, it is not merely so. If the mind, in its capacity of formal reasoning, obeys, whether consciously or unconsciously, mathematical laws, it claims through its other capacities of sentiment and action, through its perception of beauty and of moral fitness, through its deep springs of emotion and affection, to hold relation to a different order of things. Even the revelation of the material universe in its boundless magnitude, and pervading order, and constancy of law, is not necessarily the most fully apprehended by him who has traced with minutest accuracy the steps of the great demonstration. And if we embrace in our survey the interest and duties of life, how little does any process of mere ratiocination enable us to comprehend the weightier questions which they present! As truly, therefore, as the cultivation of the mathematical or deductive faculty is a part of intellectual discipline, it is only a part."

Set Theory

The beginnings of mathematical logic went pretty much hand in hand with the nineteenth century development of *set theory* – particularly the theory of *infinite sets* founded by the remarkable mathematician *Georg Cantor*. Before discussing infinite sets, we must first look at some of the basic theory of sets in general.

A set is any collection of objects whatsoever. The basic notion of set theory is that of *membership*. A set A is a bunch of things, and to say that an object x is a *member* of A, or an *element* of A, or that x *belongs* to A, or that A *contains* x, is to say that x is one of those things. For example, if S is the set of all positive integers from 1 to 10, then the number 7 is a member of A (so is 4), but 12 is not a member of A. The standard notation for membership is the symbol $\in$ (epsilon), and the expression "x is a member of A" is abbreviated "$x \in A$".

A set A is said to be a *subset* of a set B if every member of A is also a member of B. Unfortunately, many beginners learning about sets confuse *subset* with *membership*. As an example of the difference, let H be the set of all human beings and let W be the set of all women. Obviously W is a *subset* of H, since every woman is also a human being. But W is hardly a *member* of H, as W is obviously not an individual human being. The symbol for subset is the so-called "inclusion" sign $\subseteq$. Thus, for any pair of sets A and B, the phrase "A is a subset of B" is abbreviated $A \subseteq B$. If A is a subset of B, then B is called a *superset* of A. Thus a superset of A is a set that contains all elements of A, and possibly other elements as well. A subset A of B is called a *proper* subset of B if A is not the whole of B, in other words, if B contains some elements that are not in A.

A set A is the same as a set B if and only if they contain exactly the same elements, in other words, if and only if each is a subset of the other. The only way two sets can be different is if one of them contains at least one element that is not in the other.

The only way a set A can *fail* to be a subset of a set B is when A contains at least one element that is not in B.

A set is called *empty* if it contains no elements at all, such as the set of all people in a theater after everyone has left. There can be only one empty set, because if A and B are empty sets, they contain exactly the same elements – namely no elements at all. Put another way, if A and B are both empty, then neither one contains any element not in the other, since neither one contains any elements at all. Thus if A and B are both empty sets, then A and B are the same set. Thus there is only one empty set, and it will be denoted in this work by the symbol "$\emptyset$".

The empty set has one characteristic which seems quite strange to those who encounter it for the first time. As a preliminary illustration, consider a club whose president says that all Frenchmen in the club wear berets. But suppose it turns out that there are no Frenchmen in the club. Should the president's statement then be regarded as true, false, or neither? More generally, given an arbitrary property P, should it be regarded as true, false, or neither, to say that all members of the empty set have property P? Here we have to make a choice, once and for all, and the choice universally agreed upon by mathematicians and logicians is that such a statement should be regarded as *true*! One reason for such a decision is this: Given any set S and any property P, the only way that P can fail to hold for all elements of S is that there be at least one element of S for which P doesn't hold. The empty set is to be regarded as no exception to the statement just made, and so the only way that P can fail to hold for all elements of the empty set is that there is at least one element of the empty set that doesn't have the property P, but that cannot be, since there is no element of the empty set at all! [As the late mathematician Paul Halmos would say, "If you don't believe that P holds for all elements of the empty set, just try to find me an element of the empty set for which P doesn't hold!"] Thus we shall henceforth regard it as true that for *any* property P, all elements of the empty set have property P. Here is another way of looking at it, which anticipates an important principle of Propositional Logic, which we study in Part II of this volume, namely the logical use of the word "implies" or "if – then."

The phrase "if – then," as it is used in classical logic, is a bit of a shock to those encountering it for the first time, and rightfully so, since it is highly questionable whether it really corresponds to the way the phrase is commonly used.

Suppose a man tells a girl, "If I get a job next summer then I will marry you." If he gets a job next summer and marries her, he has clearly kept his word. If he gets a job and doesn't marry her, he has obviously broken his word. Now, suppose he doesn't get a job but marries her anyway. I doubt that anyone would say that he has broken his word! And so in this case too, we will say that he has kept his word. The *crucial* case is that he neither gets a job nor marries her. What would you say about this case – has he kept his word? Broken his word? Or neither? Suppose the girl complains, "You said that if you got a job you would marry me, and you didn't get any job, nor will you marry me!" The man could rightfully say, "I haven't broken my word! I never said that I will marry you – all I said is that *if* I get a job, *then* I will marry you. Since If didn't get a job, then I have not broken my word."

Well, as I said, I believe that you would not be uncomfortable with saying that he has not broken his word in this case, but I imagine many of you would be uncomfortable with saying that he has *kept his word.*

Well, we want all statements of the form "if – then" to be either true or false, regardless of whether the "if – part" or the "then-part" is true or false. Under this rule, since we have decided that the man did not break his word, we have no option but to say that he has kept his word, strange as it may seem!

Thus in classical logic, for any pair of propositions p and q, the statement "if p, then q" (also stated "p implies q") is to be regarded as false only if p is true and q is false. In other words, "p implies q" is synonymous with "it is not the case that p is true and q is false," or equivalently, "either p is false or p and q are both true," which is also equivalent to "p is false or q is true."

This type of implication is more specifically called *material implication*, and it does have the strange property that a false proposition implies any proposition! For example, the statement "If Paris is the capital of England, then $2+2=5$" is to be regarded as true!

I must tell you an amusing incident: Someone once asked Bertrand Russell, "You say that a false proposition implies any proposition. For example, from the statement $2+2=5$, could you prove that you are the Pope?" Russell replied, "Yes," and gave the following proof. "Suppose $2+2=5$. We also know that $2+2=4$, from which it follows that $5=4$. Subtracting 3 from both sides of the equation, it follows that $2=1$. Now, the Pope and I are two. Since two equals one, then the Pope and I are one! Therefore, I am the Pope".

Material implication, with all its oddity, really has its advantages, which I would like to illustrate as follows. Suppose I take a card from a deck and put it face down on the table and say, "If the card is the Queen of Spades, then it is black. Do you agree?" Surely you would agree. Then I turn the card over, and it is a red card – say the Jack of Diamonds. Would you then say that you were wrong regarding my statement as true? My case rests!

Now, how is all this about implication relevant to the statement that any property P holds for all elements of the empty set? Well, to say of a given set S and a property P that all elements of S have property P, is to say that for every element x, *if* x is in S, then x has property P. In particular, to say that all elements of the empty set $\emptyset$ have property P is to say that for any element x, *if* x is in $\emptyset$, *then* x has the property P. Well, for any x, it is false that x is in $\emptyset$, and since a false proposition implies any proposition, it is true that if x is in $\emptyset$, then x has the property P. Thus for all x, if $x \in \emptyset$, then $P(x)$, which means that P holds for all elements of $\emptyset$.

Problem 2. Is the empty set a subset of every set?

Finite sets are often displayed by enclosing the names of their elements in curly brackets – for example $\{2, 5, 16\}$ is the set whose elements are the three numbers, 2, 5, and 16. Sometimes the empty set is denoted $\{\ \}$, and I will sometimes use this notation in contexts in which we are describing members of a set by listing them inside curly brackets.

Boolean Operations on Sets

Unions

For any pair of sets A and B, by the union of A and B, denoted $A \cup B$, is meant the set of all things that belong either to A or to B, or to both. For example, if P is the set of all non-negative integers (i.e. the positive integers and zero) and N is the set of all negative integers and I is the set of all integers, then $P \cup N = I$.

Or, for another example, $\{1, 3, 7, 18\} \cup \{2, 3, 7, 24\} = \{1, 2, 3, 7, 18, 24\}$.

Problem 3. Which, if any, of the following statements are true?

(1) If $A \cup B = B$, then $B \subseteq A$.
(2) If $A \cup B = B$, then $A \subseteq B$.
(3) If $A \subseteq B$, then $A \cup B = B$.
(4) If $A \subseteq B$, then $A \cup B = A$.

We can think of $A \cup B$ as the result of adding the elements of A to the set B, or what is the same thing, adding the elements of B to A. Thus $A \cup B = B \cup A$. It is also obvious that for any three sets A, B and C, $A \cup (B \cup C) = (A \cup B) \cup C$ – if we add the elements of A to the set $B \cup C$, we get the same set as when we add the elements of $A \cup B$ to the set C. It is equally obvious that $A \cup A = A$. Also obvious is the fact that $A \cup \emptyset = A$ (we recall that $\emptyset$ is the empty set).

Intersections

For any pair of sets A and B, by their *intersection* – denoted $A \cap B$ – is meant the set of all elements that are in both A and B. For example, suppose $A = \{2, 5, 18, 20\}$ and $B = \{2, 4, 18, 25\}$. Then $A \cap B = \{2, 18\}$, since 2 and 18 are the only numbers common to both A and B. The following facts are obvious:

(a) $A \cap A = A$;
(b) $A \cap B = B \cap A$;
(c) $A \cap (B \cap C) = (A \cap B) \cap C$;
(d) $A \cap \emptyset = \emptyset$.

Problem 4. Which of the following statements are true?

(1) If $A \cap B = B$, then $B \subseteq A$.
(2) If $A \cap B = B$, then $A \subseteq B$.
(3) If $A \subseteq B$, then $A \cap B = B$.
(4) If $A \subseteq B$, then $A \cap B = A$.

Problem 5. Suppose A and B are sets such that $A \cap B = A \cup B$. Does it necessarily follow that A and B must be the same set?

Complementation

We now consider as fixed for the discussion a set I, which we will call *the universe of discourse*. What the set I is will vary from one application to another. For example, in plane geometry, I could be the set of all points on a plane. For number theory, the set I could be the set of all whole numbers. In applications to sociology the set I could be the set of all people. In the discussion of Boole's *general* theory of sets we are now embarking upon, the set I is a completely arbitrary set, and we will be considering all subsets of I.

For any subset A of I, by its *complement* (relative to I, which will be understood) is meant the set of all elements of I that are *not* in A. For example, if I is the set of all whole numbers, and E is the set of all even numbers, then the complement of E is the set of all odd numbers. The complement of a set A is denoted A', or sometimes $\overline{A}$ or $\widetilde{A}$.

It is obvious that A'' (the complement of the complement of A) is A itself.

Problem 6. Which, if either, of the following statements is true?

(1) If $A \subseteq B$, then $A' \subseteq B'$.
(2) If $A \subseteq B$, then $B' \subseteq A'$.

The operations of union, intersection and complementation are the fundamental Boolean operations on sets. Other operations are definable by iterating these fundamental operations. For example, the set denoted $A - B$ [the so-called *difference* of A and B], which is the set of all elements of A that are not in B, can be defined in terms of the three fundamental operations, since $A - B = A \cap B'$.

Venn Diagrams

Boolean operations on sets can be graphically illustrated by what are known as *Venn diagrams*, in which the universe of discourse I is represented by the set of all points in the interior of a square, subsets A, B, C etc. of I are represented by circles within the square, and Boolean operations are represented by shading appropriate portions of the circles. For example,

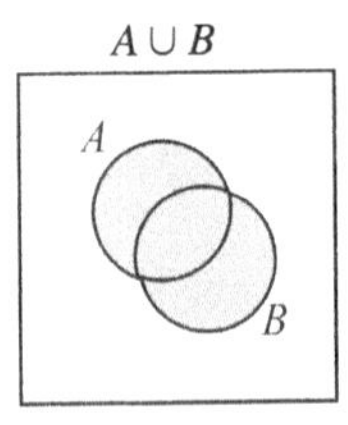

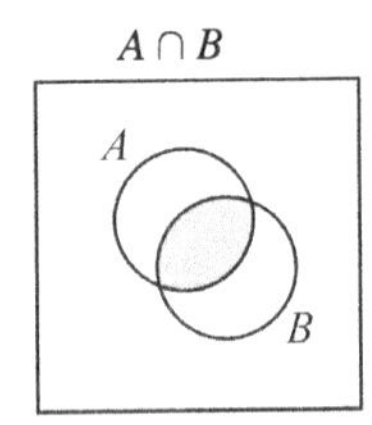

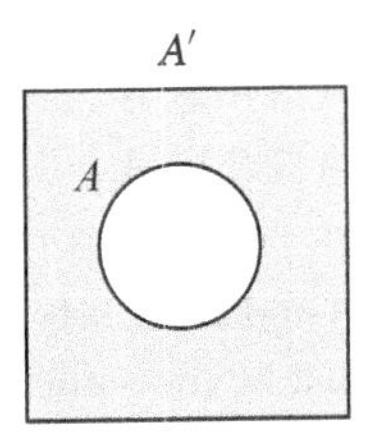

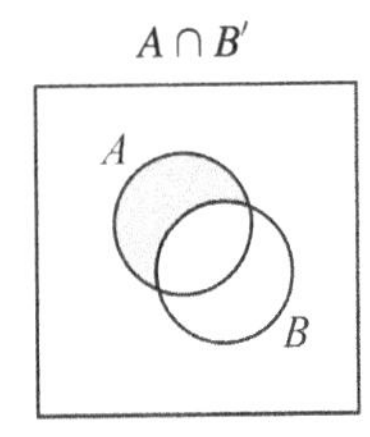

Boolean Equations

We will use the capital letters A, B, C, D, E, with or without subscripts, to stand for arbitrary sets (just as in algebra we use the lower-case letters x, y, z to stand for arbitrary numbers). We call these capital letters (with or without subscripts) *set variables*. By a *term* we mean any expression constructed according to the following rules:

(1) Each set variable standing alone is a term.
(2) For any terms t_1 and t_2, the expressions $(t_1 \cup t_2)$, $(t_1 \cap t_2)$, and t_1' are again terms.

Examples of terms are $A \cup (B \cap C')$, and $A' \cup (B \cap A'')$, and $(A \cup B)'$.

It is necessary to use parentheses to avoid ambiguity. For example, suppose we didn't use parentheses in writing the expression $A \cap B \cup C$. It is not possible to tell whether this means the union of $A \cap B$ with C or the intersection of A with $B \cup C$. If we mean the former, we should write expression $(A \cap B) \cup C$, if the latter, we should write expression $A \cap (B \cup C)$. Sometimes parentheses can be deleted if no ambiguity can result. For example, outer parentheses can be dropped, as we did above by writing $(A \cap B) \cup C$ instead of $((A \cap B) \cup C)$.

By a Boolean *equation* we mean an expression of the form $t_1 = t_2$, where t_1 and t_2 are Boolean terms. As some examples, consider:

(1) $A \cup B = A \cap B$
(2) $A' = B$
(3) $A \cup B = B \cup A$
(4) $A \cup B' = A' \cup B'$
(5) $(A \cup B)' = (A \cup B) \cap C$
(6) $A \cup (B \cap C) = (A \cup B)' \cup (C \cap (A \cap B))$

A Boolean equation is called *valid* if it is true no matter what sets the set variables represent. For example, (3) is valid, since for *any* pair of sets A and B, it is true that $A \cup B = B \cup A$. None of the other five equations above are valid.

Testing Boolean Equations

Suppose we wish to test whether or not a given Boolean equation is valid. Is there a systematic way of going about it, or is ingenuity required? The answer is that it can be done systematically. One well-known way is by use of Venn diagrams, but I have found another way [Smullyan, 2007], the method of *indexing*, to which we now turn.

As a simple starter, let A and B be subsets of I.

I (the interior of the square):

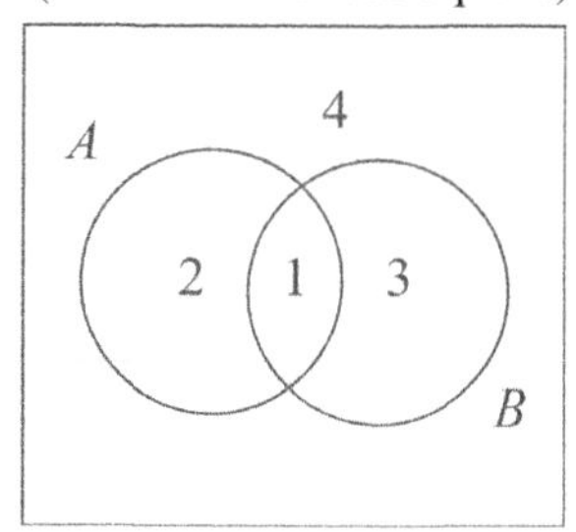

In this diagram I is divided up into the four sets of $A \cap B$, $A \cap B'$, $A' \cap B$ and $A' \cap B'$, which we have labeled (indexed) by the numbers 1, 2, 3 and 4 respectively. Every element x of I belongs to one of these four regions. Let us call these regions the *basic* regions.

Now, let us identify any union of the basic regions with its set of indices. So for example,

$$A = (1, 2)$$
$$B = (1, 3)$$

Actually, of course, we could write $A = 1 \cup 2$ and $B = 1 \cup 3$, but we are dropping the union sign to make the indices easier to focus on.

Then $A \cup B = (1, 2, 3)$; and $A \cap B = (1)$, since (1) is the only region common to A and B. Also, $A' = (3, 4)$, and since there is nothing in common between $(1, 2)$ and $(3, 4)$, then $A \cap A' = \emptyset$, i.e. $A \cap A'$ is the empty set. Also, $A \cup A' = (1, 2) \cup (3, 4) = (1, 2, 3, 4)$. Thus $A \cup A' = I$.

Now, suppose we wish to verify the De Morgan law, $(A \cup B)' = A' \cap B'$ by the method of indexing. The idea is to find first the set of indices of $(A \cup B)'$ and then the set of indices of $A' \cap B'$ and see if the two sets are the same.

$A \cup B = (1, 2, 3)$	$A' = (3, 4)$ and $B' = (2, 4)$.
Hence, $(A \cup B)' = (4)$.	Hence, $A' \cap B' = (4)$.

Thus, (4) is the set of indices of both $(A \cup B)'$ and $A' \cap B'$; hence $(A \cup B)' = A' \cap B'$.

Let's now try an equation with three sets A, B and C. These three sets divide I into eight basic regions, as the following figure shows:

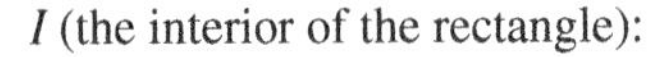

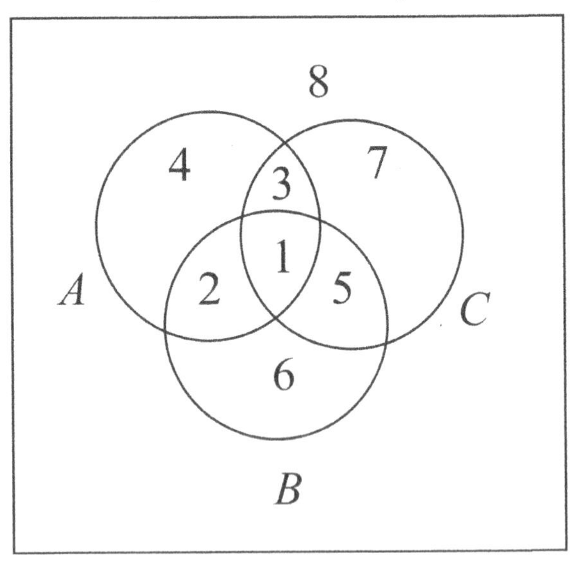

Thus:

$$A = (1, 2, 3, 4)$$
$$B = (1, 2, 5, 6)$$
$$C = (1, 3, 5, 7)$$

Suppose we wish to show that $A \cup (B \cap C) = (A \cup B) \cap (A \cup C)$. Again we reduce each side of the equality sign to its set of indices and see if the two sets are the same.

$$(B \cap C) = (1, 5) \qquad A \cup B = (1, 2, 3, 4, 5, 6)$$
$$A \cup (B \cap C) = (1, 2, 3, 4, 5) \qquad (A \cup C) = (1, 2, 3, 4, 5, 7)$$
$$(A \cup B) \cap (A \cup C) = (1, 2, 3, 4, 5)$$

Thus both sides of the set equation reduce to $(1, 2, 3, 4, 5)$ and we have won the case.

Now let's try the equation $A \cap (B \cup C) = (A \cap B) \cup (A \cap C)$.

$$(B \cup C) = (1, 2, 3, 5, 6, 7) \qquad (A \cap B) = (1, 2)$$
$$A \cap (B \cup C) = (1, 2, 3) \qquad (A \cap C) = (1, 3)$$
$$(A \cap B) \cup (A \cap C) = (1, 2, 3)$$

Thus the equation reduces to $(1, 2, 3) = (1, 2, 3)$ and we see the equation is valid.

What do we do if we have more than three unknowns, say A, B, C and D? Well, we can no longer draw circles, but still the four sets divide I into 16 basic regions, and we can number them in such a way that:

$$A = (1, 2, 3, 4, 5, 6, 7, 8)$$
$$B = (1, 2, 3, 4, 9, 10, 11, 12)$$
$$C = (1, 2, 5, 6, 9, 10, 13, 14)$$
$$D = (1, 3, 5, 7, 9, 11, 13, 15)$$

Once we number the basic regions so that we know what our original sets are in terms of unions of basic regions, we can operate as we have seen with the set of indices for these four sets.

For five unknowns, A, B, C, D and E we have 32 basic regions. In general, for any n equal or greater than 2, the sets $A_1, A_2, \ldots, A_n$ divide I into 2^n basic regions, and we can assign an index to each of these regions, say by taking A_1 to be the first half of the integers $1, 2, \ldots 2^n$, then taking A_2 to be every other quarter (starting with the first), A_3 to be every other eighth, and so forth. For example, for $n = 5$, we take:

$$A_1 = (1 - 16)$$
$$A_2 = (1 - 8, 17 - 24)$$
$$A_3 = (1 - 4, 9 - 12, 17 - 20, 25 - 28)$$
$$A_4 = (1, 2, 5, 6, 9, 10, 13, 14, 17, 18, 21, 22, 25, 26, 29, 30)$$
$$A_5 = (\text{all odd numbers from 1 to 31})$$

Exercise 1. By the method of indexing, prove the following Boolean equation to be valid:

$$(A \cup B)' \cap C = (C \cap A') \cup (C \cap B').$$

Some other Boolean operations are:
$A \to B$, which is $A' \cup B$, and
$A \equiv B$, which is $(A \cap B) \cup (A' \cap B')$.

Exercise 2. By the method of indexing, prove that the following equations are valid.

(1) $A \to B = (A - B)'$
(2) $A \equiv B = ((A \to B) \cap (B \to A))$
(3) $(A \cap (A - B)') = \emptyset$
(4) $A \equiv (A \to B)) = A \cap B$

The Boolean theory of sets is but the beginning of the field known as *Set Theory*. The deeper aspects of the subject concern *infinite* sets, which is the subject of the next chapter.

Solutions to the Problems of Chapter 1

1. *Logically* they say the same thing, namely that no food is both good and cheap, but psychologically they convey quite different images. The statement "good food is not cheap" tends to create the image of good expensive food, whereas "cheap food is not good" makes on think of cheap rotten food.
2. It is true that the empty set is a subset of every set S, because the property of being an element of S (like any other property) holds for all elements of the empty set, as we have seen. Thus every element of the empty set is an element of S, which means that the empty set is a subset of S.
3. It is (2) and (3) that are the true statements.

   Proof of (2). To begin with, it is obvious that for any pair of sets A and B, the set A is a subset of $A \cup B$. Also, $B \subseteq A \cup B$. Now, suppose $A \cup B = B$, as assumed in (2). Since A is a subset of $A \cup B$, it is a subset of B, because B is the same set as $A \cup B$. Thus $A \subseteq B$.

   Proof of (3). Suppose $A \subseteq B$. It is also true that $B \subseteq B$ (obviously!). Thus A and B are both subsets of B, and hence so is their union $A \cup B$.
4. It is (1) and (4) that are the correct statements.

   Proof of (1). Suppose $A \cap B = B$. Let x be any element of B. Since $B = A \cap B$, then x is an element of $A \cap B$, which means that x belongs to both A and B, so that, in particular, $x \in A$. This proves that any element x of B is also an element of A, and so $B \subseteq A$.

   Proof of (4). Suppose $A \subseteq B$. Let x be any element of A. Since $A \subseteq B$, then also $x \in B$. Thus $x \in A$ and $x \in B$, so that $x \in A \cap B$. Thus any element of A is also an element of $A \cap B$, which means that $A \subseteq A \cap B$. Also, $A \cap B$ is obviously a subset of A. Thus A and $A \cap B$ are subsets of each other, and so $A \cap B = A$.
5. Yes, it is true that A must equal B when $A \cap B = A \cup B$. For suppose A were different from B. Then one of the two sets, say A, contains an element x not in the other set, B. Since $x \in A$, then $x \in A \cup B$. Since $x \notin B$, then $x \notin A \cap B$. Thus $x \in A \cup B$, but not in $A \cap B$. This proves that if A is different from B, then $A \cup B$ is different from $A \cap B$. But we are given that $A \cup B$ is not different from $A \cap B$; hence A cannot be different from B. Thus $A = B$.
6. It is (2) that is true.

   Proof of (2). Suppose $A \subseteq B$. Then any element x of B' is not in B, hence not in A, hence must be in A'. Thus every element of B' is also in A', which means $B' \subseteq A'$.

2
Infinite Sets

A complete mathematical revolution occurred in the late nineteenth century with Georg Cantor's theory of infinite sets. Just what does it mean for a set to be finite or infinite? The basic idea behind this is that of a 1-1 *(one to one) correspondence.*

Suppose that in a certain theater, every seat is taken and no one is standing and no one is sitting on anyone's lap. Then without having to count the number of people or the number of seats, we know that the two numbers must be the same, because the set of people is in a 1-1 correspondence with the set of seats – namely, we see each person as in correspondence with the seat upon which he or she is sitting. More generally, by *a 1-1 correspondence between a set A and a set B* is meant an assignment which pairs each element x of A with one and only one element y in B, while at the same time each element y of B is paired with one and only one element x in A. Thus the elements in the two sets are "paired up," with each element in either set having a single corresponding partner in the other set.

We are all familiar with the so-called *natural numbers* – the whole numbers $0, 1, 2, \ldots, n, \ldots$. The natural numbers thus consist of 0 together with the positive integers (themselves sometimes called the whole numbers). [I'll tell you something funny: When once in conversation, I used the term "natural numbers," someone looked puzzled and asked me, "Can you give me an example of an *unnatural* number?"]

For any *positive* natural number n, to say that a set A has exactly n elements is to say that A can be put into a 1-1 correspondence with the set of positive whole numbers from 1 to n. For example, to say that there are five fingers on my left hand is to say that I can put the set of fingers on my left hand into a 1-1 correspondence with the set of whole numbers from 1 to 5, for example, my thumb to 1, my index finger to 2, my middle finger to 3, the next finger to 4, and my pinky to 5. This act of putting a set into a 1-1 correspondence with the set of positive integers from 1 to some positive integer n act has a popular name. It is commonly called *counting*. Yes, that's exactly what counting is. We also say that a set has *zero* elements if it has no elements at all, in other words if it is the empty set.

Now that we have defined what it means for a set to have n elements, we can now define a set to be *finite* if there is a natural number n such that the set has (exactly) n elements. Then we define a set to be *infinite* if it is not finite. An obvious example of an infinite set is the set N of *all* natural numbers. Since there is no greatest natural number, there cannot be any natural number n such that N has exactly n elements. For any natural number n, the set N obviously has more than n elements.

Problem 1. If a set A has exactly three elements, how many subsets of A are there (including A itself and the empty set)? What about a 5-element set? In general, if a set A has n elements, then in terms of n, how many subsets of A are there?

Sizes of Infinite Sets

How should one define what it means for two nonempty sets A and B to be of the same numerical size? The obvious and correct answer is that A can be put into a 1-1 correspondence with B.

Given two nonempty sets A and B, how should one define what it means for the set A to be numerically *smaller* than the set B or for the set B to be numerically *larger* than the set A? An obvious guess is that one should define A to be (numerically) *smaller* than B if A can be put into a 1-1 correspondence with a *proper* subset of B (a subset of B that is not the whole of B). This definition would be fine for *finite sets*, but is unfortunately useless for infinite sets since it can happen that two infinite sets A and B can be found such that on the one hand, A can be put into a 1-1 correspondence with a proper subset of B, while on the other hand B can also be put into a 1-1 correspondence with a proper subset of A. But this would make it true that both A is smaller than B and B is smaller that A, which we certainly don't want. For example, let E be the set of even natural numbers, and let O be the set of all odd natural numbers. We can put E into a 1-1 correspondence with a *proper* subset of O by pairing each even number n with, say, $x+5$. Thus 0 is paired with 5, 2 with 7, 4 with 9, etc. Thus E is in 1-1 correspondence with the set of all odd numbers other than 1 and 3, which is a *proper* subset of O. On the other hand, we can put O into a 1-1 correspondence with a proper subset of E by pairing each odd number x with $x+3$. Thus 1 corresponds with 4, 3 with 6, 5 with 8, etc. Thus O is 1-1 correspondence with the proper subset of E consisting of all even numbers other than 2 and 4.

No, the above definition won't work for infinite sets. The correct definition of a nonempty set A being smaller than a nonempty set B is that A can be put into a 1-1 correspondence with a subset of B, but A cannot be put into 1-1 correspondence with the whole of B. Thus if A is smaller than B, then *every* 1-1 correspondence with a *proper* subset of B will leave out some elements of B.

A curious thing about infinite sets is that each of them can be put into a 1-1 correspondence with a *proper* subset of itself! We will later prove that about infinite sets in general, but for now, we will see that it is true for the set N of natural numbers. N can be matched with just the set of positive integers as follows:

Natural Numbers	0	1	2	3	...	n	...
	$\updownarrow$	$\updownarrow$	$\updownarrow$	$\updownarrow$		$\updownarrow$	
Positive Integers	1	2	3	4	...	$n+1$	...

Actually, N can be matched with just the set of even natural numbers as follows:

Natural Numbers	0	1	2	3	...	n	...
	$\updownarrow$	$\updownarrow$	$\updownarrow$	$\updownarrow$		$\updownarrow$	
Even Natural Numbers	0	2	4	6	...	$2n$	...

Perhaps even more startling is the fact observed by Galileo in 1630 that the set N of natural numbers can be matched with the set of perfect squares $(0, 1, 4, 9, 16, 25, \ldots)$, which gets sparser the further out we go:

Natural Numbers	0	1	2	3	...	n	...
	$\updownarrow$	$\updownarrow$	$\updownarrow$	$\updownarrow$		$\updownarrow$	
Perfect Squares	0	1	4	9	...	n^2	...

Isn't there some sense in which the set of perfect squares is smaller than the set N of all natural numbers? Yes, and that is in the sense that the set of perfect squares form only a *proper* subset of N. Nevertheless the set of perfect squares is *numerically* of the same size as N.

Denumerable Sets

The first basic question about infinite sets considered by Cantor is whether all infinite sets are of the same size, or whether they come in different sizes. Would you care to hazard a guess at this point?

A set is called *denumerable*, or *denumerably infinite*, if it can be put into a 1-1 correspondence with the set of positive integers, or what is the same thing, when it can be put into a 1-1 correspondence with the set of natural numbers.

A 1-1 correspondence between a set A and the set of positive integers is called an *enumeration* of A. An enumeration of A can be thought of as an infinite list $a_1, a_2, \ldots, a_n, \ldots$, where each n of N is paired with the element a_n of A. The number n is sometimes called the *index* of a_n in the enumeration.

The big question now if whether every infinite set is denumerable. As I understand it, for several years Cantor thought that all infinite sets were denumerable, but one day he realized that the opposite was the case! What he did (for several years, I am told) was to examine various sets that on the surface appeared to be non-denumerable, but in each case he found a clever way to enumerate the set. I like to illustrate these clever enumerations in the following manner.

Imagine that you and I are immortal. I write down on a piece of paper some positive integer, and tell you that each day you have one and only one guess as to what the number is. I offer you a grand prize if you ever guess my number. Is there a strategy you can take that will guarantee that you will get the prize sooner or later? Obviously there is: On the first day, you ask if the number is 1; on the second day, whether it is 2, and so forth. If my number is n, then you will win the prize on the nth day.

Next, I make the problem a wee bit harder. This time I write down either a positive whole number $1, 2, 3, \ldots, n, \ldots$ or a negative whole number $-1, -2, -3, \ldots, -n, \ldots$. Again you are allowed only one guess a day.

Problem 2. What strategy would enable you to be sure that you will certainly win the prize sooner or later?

My next test if definitely harder. This time I will write down two whole numbers (possibly the same one twice). Again you have only one guess per day, and you must guess both of the numbers on the same day. You are not allowed to guess one of them one day, and the other on a different day. This might seem hopeless, since there are infinitely many possibilities for the first number I write down, and for each of those possibilities, there are infinitely many possibilities for the second number. Thus it seems there is no sure way for you to win. But there is a way!

Problem 3. What strategy would work?

In the next test, I write down a fraction x/y, where x and y are positive integers. Again on each day you have one and only one guess as to what fraction I wrote.

Problem 4. What strategy would enable you to be sure of naming my fractions on some day or other?

We have now seen that the set of all fractions is denumerable, i.e. the same size as the set of all positive integers, since we can pair each number n with the fraction named on Day n according to the scheme given in the solution to the above problem. Cantor's discovery that the set of fractions is denumerable came as quite a shock to the mathematical world!

In my next test I write down a finite set of positive integers. I don't tell you how many integers I wrote down, nor what the largest of the integers happens to be. Again you have one guess per day as to the set of integers that I wrote.

Problem 5. What strategy would enable you to be sure of finding my set sooner or later?

Cantor's Great Discovery

We have just seen that the set of all *finite* sets of positive integers is denumerable. What about the set of *all* sets of integers, whether finite or infinite? Is that set denumerable? Cantor's great discovery is that the set of all sets of positive integers is *not* denumerable! We now turn to the proof of this crucial fact, which started the whole mathematical field known as *set theory.*

To illustrate the proof, let us consider a book with denumerably many pages – Page 0, Page 1, Page 2, ..., Page n, ... On each page is written down a description of a set of natural numbers. If *every* set of natural numbers is listed somewhere in the book, then the book wins a grand prize. But without even looking at the contents of the book, one can describe a set of natural numbers that cannot possibly be listed on any page!

Problem 6. What set could that be? What set is such that it cannot be listed on any page? (Hint: Call a number n *extraordinary* if n is a member of the set described on Page n, and *ordinary* if it is not a member of the set listed on Page n. Is it possible for the set of all extraordinary numbers to be listed on some page? Is it possible for the set of all ordinary numbers to be listed on some page?)

For any set A, the set of all subsets of A is called the *power set of* A, and is denoted $\mathcal{P}(A)$. Thus in our solution to Problem 6 what we have shown is that N cannot be put into a 1-1 correspondence with its power set $\mathcal{P}(N)$. However, N can certainly be put into a 1-1 correspondence with a *subset* of $\mathcal{P}(N)$.

Problem 7. Why is it true that N can be put into a 1-1 correspondence with a subset of $\mathcal{P}(N)$?

Since N cannot be put into a 1-1 correspondence with $\mathcal{P}(N)$, but can be put into a 1-1 correspondence with a subset of $\mathcal{P}(N)$, then, by definition, the set $\mathcal{P}(N)$ is numerically larger than N! This is but a special case of Cantor's theorem, which is

Cantor's Theorem. For any set A, the power set $\mathcal{P}(A)$ is numerically larger than A.

Problem 8. Prove Cantor's theorem. [The proof is hardly different than for the special case of N.]

Continuum Problem

We have now proven Cantor's famous result that for any set A, the power set $\mathcal{P}(A)$ is numerically larger than A. In particular, $\mathcal{P}(N)$ is larger than N. But then $\mathcal{P}(\mathcal{P}(N))$ is still larger than $\mathcal{P}(N)$, and larger still is $\mathcal{P}(\mathcal{P}(\mathcal{P}(N)))$, and so forth ad infinitum.

Thus there are infinitely many different sizes for infinite sets. It is known that the set $\mathcal{P}(N)$ is the same size as the set of all points on a straight line, and accordingly the set $\mathcal{P}(N)$ is known as *the continuum.* Now comes the most interesting question: Is there a set S intermediate in size between N and $\mathcal{P}(N)$? In other words, is there a set S that is larger than N and smaller than $\mathcal{P}(N)$? Or is it that $\mathcal{P}(N)$ is of the next size larger than N? Cantor conjectured that there was no intermediate size, and this conjecture is known as *the continuum hypothesis.* More generally, Cantor conjectured that for every infinite set A, there is no set whose size is intermediate between that of A and that of $\mathcal{P}(A)$ and this conjecture is known as the *generalized continuum hypothesis.*

To this day, no one knows whether the continuum hypothesis, in either form, is true or false. Some (including myself) regard this as the *grand* unsolved problem of mathematics. This much, however, is known: In the late 1930's Kurt Gödel showed that in the most powerful mathematical system known today the continuum hypothesis is not disprovable. Then, in the 1960's, Paul Cohen showed that the continuum hypothesis is not provable in the same system. Both the continuum hypothesis and its negation – each is consistent with the axioms of present-day set theory. Thus the present-day axioms of set theory are insufficient to settle the problem. Now, there are those mathematicians known as *formalists* who regard this as a sign that the continuum hypothesis is neither true nor false in its own right, but depends on what axiom system is used. Then there are others, called *Platonists* (and this includes the present author) who believe, quite independent of any axiom system, that the generalized continuum hypothesis is either true or false in its own right, but we simply don't know which is the case. It is interesting that Gödel himself, who proved that the continuum hypothesis is consistent with the axioms of set theory, nevertheless believed that when more is known about sets, the continuum hypothesis will be seen to be false.

Problem 9. Is the union of two denumerable sets necessarily denumerable?

Problem 10. We already know from the solution to Problem 5 that the set of all finite sets of natural numbers is denumerable. (We actually proved this in the solution to Problem 5 for the positive integers, but the proof is no different for the natural numbers.) What about the set of all infinite sets of natural numbers? Is that set denumerable or not?

Problem 11. Consider a denumerable sequence $D_1, D_2, \ldots, D_n, \ldots$, of denumerable sets, and let S be their union, that is, the set of all elements x that belong to at least one of the sets $D_1, D_2, \ldots, D_n, \ldots$. Is the set S denumerable or not?

Problem 12. Given a denumerable set D, consider the set of all *finite sequences* of elements of D. Is that set denumerable?

Problem 13. Consider the set of all *infinite sequences* of 1's and 0's. Prove that set to be the same size as $\mathcal{P}(N)$.

Exercise. Prove that if D is a denumerable set, then any infinite subset of D must also be denumerable.

Discussion

Here is a significant difference between denumerable and non-denumerable sets: Imagine that you and I are immortal. I give you a check that says on it "Payable at some bank." If there are only finitely many banks in the universe, then of course you will get your money one day. Even if there are denumerably many banks, Bank 1, Bank 2, . . . , Bank n, . . . , you can be sure of collecting one day, though you have no idea of how long it will take. However, if there are non-denumerably many banks, you chances of collecting are infinitely small!

Problem 14. Prove that every infinite set has a denumerable subset.

Problem 15. Prove that every infinite set, even a non-denumerable one, can be put into a 1-1 correspondence with a proper subset of itself.

The Bernstein Schroeder Theorem

Consider an infinite set M of men and an infinite set W of women such that every man loves one and only one of the women (but some of the women may be unloved), and no two men love the same woman. Also, each woman loves one and only one man (but not necessarily a man who loves her), and no two women love the same man (but some of the men may be unloved).

The problem is to prove that all the men can be married to all the women in a monogamous way, and moreover in such a way that in each of the married couples, either the man loves his wife, or the wife loves her husband (but unfortunately there is no way to guarantee both; indeed if some man or woman in initially unloved, then it is obviously impossible to guarantee both).

How can this be done? I'll give you a hint: We divide all the people into three groups in the following manner. Given a person x, we define a *path* starting with x as follows: If x is unloved, that's the end of the path. If x is loved, let x_1 be the one of the opposite sex who loves x. If x_1 is unloved, that's the end of the path; otherwise, let x_2 be the lover of x_1. We continue on and on in this way, and there are three possible outcomes for a given x. Either the path ends in some unloved man, in which case we will say that x belongs to Group 1; or the path ends in some unloved woman, in which case we will say that x belongs to Group 2; or the path goes on forever, in which case we put x in Group 3.

Problem 16. With this hint, finish the proof.

The mathematical content of the above problem is:

The Bernstein Schroeder Theorem. For any pair of infinite sets A and B, if A can be put into a 1-1 correspondence with a part of B (i.e. a subset of B) and B can be put into a 1-1 correspondence with a part of A, then the whole of A can be put into a 1-1 correspondence with the whole of B.

Moreover, if C_1 is a 1-1 correspondence between A and a subset of B, and C_2 is a 1-1 correspondence between B and a subset of A, then there is a 1-1 correspondence C between the whole of A and the whole of B such that for any x in A, and y in B, if x is paired with y under C, then either x was originally paired with y under C_1 or y was originally paired with x under C_2.

Problem 17. Suppose A is an infinite set of the same size as a subset of B, where B is a subset of A. Is A necessarily of the same size as B?

Solutions to the Problems of Chapter 2

1. For any positive integer n, let I_n be the set $\{1, \ldots, n\}$ of all numbers from 1 to n inclusive. Let us note that the number of subsets of I_{n+1} is twice that of I_n, because the subsets of I_{n+1} are the subsets of I_n, together with those same subsets, each one with the number $n+1$ adjoined to it. The notation for adjoining an element x to a set S is $S \cup \{x\}$. Thus if $S_1, \ldots, S_k$ are the subsets of I_n, then the subsets of I_{n+1} are the $2k$ sets

$$S_1, \ldots, S_k, \quad S_1 \cup \{n+1\}, \ldots, S_k \cup \{n+1\}.$$

Thus we see that the number of subsets of I_{n+1} is twice that of I_n.

Now, there are 2 subsets of $\{1\}$, namely $\{1\}$ itself and the empty set $\{\ \}$. If we adjoin the number 2 to each of these sets, we have the two sets $\{2\}$ and $\{1,2\}$, and so there are the 4 subsets of $\{1,2\}$, namely $\{\ \}$, $\{1\}$, $\{2\}$, $\{1,2\}$. Thus the number of subsets of I_2 is 2^2. If we take these 4 sets together with each of these sets with 3 adjoined, we get $2^3 (=8)$ subsets of I_3, namely

$$\{\,\}, \{1\}, \{2\}, \{1,2\}, \{3\}, \{1,3\}, \{2,3\}, \{1,2,3\},$$

and so forth. Thus for any positive n, there are 2^n subsets of an n-element set.

There is, of course, just 1 $(=2^0)$ subset of the empty set, namely the empty set itself.

2. One person incorrectly proposed the following solution: "First go through all the positive integers, then go through the negative ones." If I have written down

a positive integer, this will work, but if I have written a negative one, he will never get to it, since there are infinitely many positive integers he must first ask about.

No, the obvious strategy is to alternate between the positives and the negatives: On the first day you ask if the number is $+1$; on the next day whether it's -1; on the next day whether it's $+2$, on the following day, whether it's -2, and so forth.

3. There is only one possibility if my highest number is 1, namely $(1, 1)$. There are 2 possibilities for the case that my highest number is 2, namely $(1, 2)$ and $(2, 2)$. In general, for each positive integer n, there are only finitely many possibilities if the highest of my two numbers is n, namely the n pairs $(1, n), (2, n), \ldots$ $(n-1, n), (n, n)$. Thus you first go through all the possibilities in which the highest number is 1, then all the possibilities in which the highest number is 2, and so forth.
4. For each positive n, there are exactly $2n - 1$ fractions in which n is the highest of the numerator and denominator, namely the fractions $\frac{1}{n}, \frac{2}{n}, \ldots,$ $\frac{n-1}{n}, \frac{n}{n}, \frac{n}{1}, \frac{n}{2}, \ldots \frac{n}{n-1}$. Thus you first go through all the fractions in which the highest of the numerator and denominator is 1, then through all the fractions in which the highest of the numerator and denominator is 2, and so forth. Note that here we mean by fraction just a positive integer over a positive integer, so that 2/3 is a different fraction from 4/6, even though they both represent the same rational number.
5. As we saw in the solution to Problem 1, for any positive integer n, there are exactly 2^{n-1} sets of positive integers whose highest number is n, namely all subsets of $\{1, \ldots, n-1\}$ with n adjoined, and as we already know, there are 2^{n-1} subsets of $\{1, \ldots, n-1\}$.

   We can thus go through all the sets whose highest number is 1, then all the sets whose highest number is 2, and so forth.
6. It is the set – call it S – of ordinary numbers that cannot be listed on any page.

   For each number n, let S_n be the set of natural numbers listed on page n. The set S of ordinary numbers must be different from every one of the sets S_0, S_1, $S_2, \ldots, S_n, \ldots,$ because, for each n, the number n must belong to one of the sets S_n or S, but not to both of them. Reason: The number n is either ordinary or extraordinary. If it is ordinary, then by definition it does not belong to S_n, but it must then belong to S, which is the set of *all* ordinary numbers. So in this case n belongs to S but not to S_n. On the other hand, if n is extraordinary, it belongs to S_n by definition of "extraordinariness," but it cannot belong to S, which contains only ordinary numbers. Thus in this case, n belongs to S_n but not to S. This proves that S is different from every S_n, since n belongs to one of the two sets S_n or S, but not to both of them.
7. We already know that the set $\mathcal{F}$ of all finite sets of natural numbers is denumerable, and that $\mathcal{F}$ is a subset of $\mathcal{P}(N)$.

Alternatively, we could pair each natural number n with the unit set $\{n\}$, the set whose only element is n.

8. Cantor's Theorem holds for all sets, finite and infinite. It was proved for finite sets in the solution to Problem 1 (and remember that both the empty set and its power set are finite, the first having zero elements, and the second having a single element, the empty set). So we will prove this theorem here only for infinite sets.

   Given an infinite set A, consider any 1-1 correspondence which pairs each element x of A with a subset of A, and call the subset of A that x is paired with S_x. Again define x to be *ordinary* if x does not belong to S_x. The set S of ordinary elements of A is different from each element S_x of $\mathcal{P}(A)$, since S contains x and S_x doesn't if x is ordinary, and S_x contains x but S doesn't if x is not ordinary. Thus A cannot be put into a 1-1 correspondence with $\mathcal{P}(A)$, but A can be put into a 1-1 correspondence with a subset of $\mathcal{P}(A)$ by pairing each element x of A with the unit set $\{x\}$.

9. Of course it is. Given an enumeration $a_1, a_2, \ldots, a_n, \ldots$ of a denumerable set A, and an enumeration $b_1, b_2, \ldots, b, \ldots$ of a denumerable set B, we can enumerate $A \cup B$ in the order $a_1, b_1, a_2, b_2, \ldots, a_n, b_n, \ldots$.

10. Since the set of all finite subsets of N is denumerable, then if the set of all infinite subsets of N were denumerable, the union of these two sets would be denumerable (as we saw in the last problem) but this union is $\mathcal{P}(N)$, which is non-denumerable. Therefore the set of all infinite subsets of N cannot be denumerable, thus must be non-denumerable.

11. Let D be the union of the denumerable sets $D_1, D_2, \ldots, D_n, \ldots$. For each n, let $d_n(1), d_n(2), \ldots, d_n(m), \ldots$ be an enumeration of the set D_n. We know that the set of fractions is denumerable, and we can enumerate the elements of D in the same order as we did the fractions, i.e. we start with all the elements $d_n(m)$, in which the highest of n and m is 1, then those in which the highest of n and m is 2, and so forth. We eliminate the duplicates that may arise at each step due to the fact that the sets D_i are not required to be mutually disjoint (although each one has denumerably many mutually distinct elements by itself).

12. To begin with, let us note that for any finite set F with, say, m elements, and any positive integer k, the set of all sequences (allowing repetitions) of length k of elements of F is a finite set; specifically, there are m^k such sequences (since there are m possible choices for the first term of the sequence, and with each choice, there are m choices for the second term, hence $m \times m = m^2$ choices for the first two terms, and with each such choice, there are m choices for the third term, hence $m^2 \times m = m^3$ choices for the first three terms, and so forth).

    Now consider a denumerable set D. We will show that the set of all finite sequences of elements of D is denumerable. Well, for each positive integer n, let $(S_k)_n$ be the set of all sequences of length k of elements of $\{a_1, \ldots, a_n\}$ where each sequence contains at least one copy of a_n, and let S_k be the set of all finite sequences of D of length k. Each set $(S_k)_n$ is finite (as shown above). Note that for every k each set $(S_k)_n$ is different from all the $(S_k)_m$ with $m \neq n$, and each S_k is different from every S_j with $j \neq k$. We can enumerate all elements of S_k by starting with the elements of $(S_k)_1$ (of which there is only one) followed by the finitely many elements of $(S_k)_2$ (in any order, and deleting repetitions), followed by those of $(S_k)_3$, and so forth. Thus each S_k is denumerable.

Since each of the sets $S_1, S_2, \dots, S_n, \dots$ is denumerable, so is their union (by Problem 11), and the union is the set of all finite sequences of elements of D.

13. We pair each infinite sequence θ of 1's and 0's with the set of all positive integers n such that the nth term of θ is 1. As examples, the sequence $(1, 0, 1, 0, 1, 0, \dots)$ is paired with the infinite set $\{1, 3, 5, 7, \dots\}$ of odd numbers. The sequence $(1, 0, 1, 1, 0, 1, 0, 0 \dots 0, 0, 0, \dots)$ is paired with the finite set $\{1, 3, 4, 6\}$.

    Obviously distinct sequences are then paired with distinct sets of natural numbers, and for each set of natural numbers A, there is one and only one sequence that is paired with A, namely the sequence in which for all n the nth term of the sequence is 1 if n is in A, and is 0 if n is not in A. Thus the pairing is a 1-1 correspondence between the set of all the sequences to the set of all sets of positive integers, which of course is the same size as the set of all sets of natural number, $\mathcal{P}(N)$, since the set of positive integers is the same size as the set of natural numbers.

14. It is obvious that if we remove an element x from an infinite set A, the resulting set $A - \{x\}$ must be infinite. For if it were finite, it would have n elements for some natural number n, so that the original set A would have had $n + 1$ elements, contrary to the assumption that A is infinite.

    Now, consider an infinite set A. A is obviously non-empty, hence we can remove an element a_1. What remains is infinite, hence we can remove another element a_2., and so forth. We thus generate a denumerable sequence $a_1, a_2, \dots, a_n, \dots$ of elements of A.

    Remark. Hidden in the above proof is a principle known as the Axiom of Choice, which we will unfortunately not be able to consider in this work.

15. Consider an infinite set A. As we have just seen, A includes a denumerable subset $D = \{d_1, d_2, \dots, d_n, \dots\}$. This set D can be put into a 1-1 correspondence with its proper subset $\{d_2, d_3, \dots, d_{n+1}, \dots\}$ by pairing each d_n with d_{n+1}. To augment this pairing of a subset of A with another subset of A to obtain a pairing from all of A to a subset of A, we let each element of A other than those in D correspond with itself. Then A is in 1-1 correspondence with $A - \{d_1\}$.

16. Obviously every unloved man is in Group 1 and every unloved woman is in Group 2. Therefore every person in Group 3 is loved. Furthermore:
    (a) For every man in Group 1, the woman he loves is also in Group 1, and every woman in Group 1 is loved by some man (since she is not in Group 2), and

this man is also in Group 1. Thus, if all the men in Group 1 marry the women they love, these women are all in Group 1 and include all the women in Group 1.

(b) Similarly, if every woman in Group 2 marries the man she loves, then all the women in Group 2 will be married to men in Group 2, and every man in Group 2 will be the husband of a woman in Group 2.

(c) Every man in Group 3 both loves a woman in Group 3 and is loved by a woman in Group 3, and every woman in Group 3 both loves a man in Group 3 and is loved by a man in Group 3. Therefore we have the choice of either marrying all the men in Group 3 to the women they love, or marrying all the women to the men they love. [Which is the better choice is a problem best left to a psychologist.] In either case, all the men and all the women in Group 3 can be married.

17. Since B is a subset of A, then it is obviously the same size as a subset of A (namely B itself), and since A is the same size as a subset of B, it follows that A is the same size as B by the Bernstein Schroeder Theorem.

3

Some Problems Arise!

The Paradoxes

Shortly after the start of Cantor's Set Theory, some paradoxes arose that threatened the validity of the whole theory of infinite sets!

One such paradox is this: Consider the set S of all sets. Its power set $\mathscr{P}(S)$ is a subset of S, because every element of $\mathscr{P}(S)$ is a set, hence an element of the set S, the set of all sets. Also, S is the same size as a subset of $\mathscr{P}(S)$ (by a proof similar to that given as the solution to Problem 7 of Chapter 2, taking S for N). Thus S is the same size as a subset of the subset $\mathscr{P}(S)$ of S, hence by Problem 17 of Chapter 2, S is the same size as $\mathscr{P}(S)$, contrary to Cantor's Theorem.

Next there is the famous Russell Paradox (discovered independently by Zermelo). Call a set *ordinary* if it is not a member of itself, and *extraordinary* if it is a member of itself. Whether or not extraordinary sets exist, ordinary sets obviously do. Let M be the set of all ordinary sets. Is M ordinary or not? Either way we get a contradiction: Suppose M is ordinary. Then M is in M, which contains *all* ordinary sets, but being in M makes M extraordinary by definition. Thus it is paradoxical to assume that M is ordinary. On the other hand, suppose M is extraordinary. This means that M is a member of itself, i.e. M is in M. But the only sets in the set M are ordinary sets. This is again a contradiction.

Russell subsequently made a popular version of this paradox. A male barber of a certain town shaved all men of the town who did not shave themselves, and only such men. Thus if a man of the town did not shave himself, the barber would shave him, but if a man of the town shaved himself, the barber would not shave him. Did the barber shave himself or didn't he? If he shaved himself, then he shaved someone who shaved himself, which he was not supposed to do. If he failed to shave himself, then he failed to shave someone who didn't shave himself, which violates the given condition that he always shaves anyone who doesn't shave himself. Thus either way we get a contradiction.

Problem 1. The solution to the barber paradox is really very simple! What is it?

I must tell you a very amusing incident about the barber paradox. I told it to a friend of mine who has an excellent sense of humor. She said, "He probably went to his brother's house in another town and shaved himself."

There are several variants of Russell's paradox that do not involve the notion of sets. For example, call an adjective *autological* if it has the property it describes, and call it *heterological* if it does not. For example, the adjective "polysyllabic" is itself polysyllabic, hence it is autological, whereas the adjective "monosyllabic" is not monosyllabic, hence is heterological. Now, what about the word "heterological?" Is it heterological or not? Either way we get a contradiction.

Then there is the *Berry* paradox. Consider the following description (where "number" means "natural number"):

THE SMALLEST NUMBER NOT DESCRIBABLE IN LESS THAN ELEVEN WORDS.

The above description uses only ten words!

Problem 2. What is the solution to the Berry paradox?

I would now like to tell you of an amusing paradox I recently thought of. Let us go back to the book of infinitely many pages, upon each of which is listed a description of a set of natural numbers. We recall that the set of ordinary numbers (numbers n that do not belong to the set listed on Page n) – this set cannot be any one of the sets listed on any page. But now, suppose that on one of the pages, say Page 13, is written "the set of all ordinary numbers." Is 13 ordinary or not? If it is ordinary, then it belongs to the set of all ordinary numbers, which is the set listed on Page 13, which makes 13 extraordinary! If 13 is extraordinary, then by definition it is a member of the set listed on page 13, which is the set of all ordinary numbers, and therefore 13 must be ordinary! Either way we get a contradiction!

Problem 3. What is the solution to that paradox?

Hypergame

In 1987, the mathematician William Zwicker came out with a lovely paradox called "Hypergame," which he later transformed into a completely new and fascinating proof of Cantor's theorem!

We consider only 2-person games. Call a game *normal* if it has to terminate in a finite number of moves. For example, tic-tac-toe is obviously normal. Chess is a normal game, if played by tournament rules. Now, here is *hypergame*: The first move of hypergame is to choose what normal game should be played. Suppose, for example, that you and I are playing hypergame and that I have the first move. Then

I must declare what normal game should be played. I might say, "Let's play chess," in which case you make the first move in chess, and we keep playing until the chess game terminates. Or, instead, I might say, "Let's play tic-tac-toe," and you then make the first move in tic-tac-toe. I can choose any *normal* game I like, but I am not allowed to choose a game that is not normal. Then the second player makes the first move in the game chosen by the first player, and the two players play that normal game, which must end sometime. That is all there is to the rules of hypergame.

The problem is this: Is hypergame normal or not? First I will prove that hypergame must be normal. For the first player must choose some normal game. This normal game terminates in n moves for some positive integer n, hence this hypergame has terminated in $n+1$ moves. Thus hypergame must be normal. Now that hypergame has been established to be normal, consider a situation in which I, as the first player, say "Let's play hypergame!" Then you might say as your first move in hypergame, "Let's play hypergame." Then I can say, "Let's play hypergame," and so on indefinitely. And so the game never terminates, which proves that hypergame is not normal. Thus hypergame is both normal and not normal. This is the paradox!

Problem 4. What is the solution?

Someone once said, "A paradox is a truth standing on its head to attract attention." Well, out of hypergame, Zwicker extracts the following remarkable alternative proof of Cantor's theorem.

Again we consider an infinite set A and a 1-1 correspondence between A and its power set $\mathcal{P}(A)$ that assigns to each element x of A a subset of A that we will denote S_x. The problem is to show that there exists a subset S of A that is different from every S_x. Cantor's set S is the set of all ordinary elements of A (all elements x such that x is not a member of S_x). Zwicker exhibited such a set S that is completely different from Cantor's set! Here is how!

By a *path* we mean any finite or infinite sequence constructed as follows: We start with any element x of A. We go to the set S_x. If S_x is empty that is the end of the path. If S_x is not empty we take some element y of S_x and then go to the set S_y. If S_y is empty that is the end of the path. If S_y is not empty we take some element z of S_y and so forth. Either we eventually hit the empty set, or the path goes on forever. We now define an element x of A to be *normal* if *all* paths starting with x are finite, i.e. all paths starting with x must terminate in the empty set.

Well, as Zwicker proved (and the proof is not hard), the set M of all *normal* elements of A cannot be S_x for any x.

Problem 5. Prove this.

As previously noted, the solution to the barber paradox, which is the popularization of Russell's paradox, is quite simple – there can be no such barber. But Russell's paradox is far more serious. Before discussing the proposed remedies, I would like to present an amusing (and perhaps annoying!) variant of my paradox of Problem 3, the paradox in which on some page of the book is written "The set of all ordinary numbers." We saw that the solution is that what was written was not a genuine description, but only a pseudo-description. Now, to make life more complicated, we consider another book with infinitely many pages, but this time, on each page is written either a genuine description or a pseudo-description. Now consider the following description:

THE SET OF ALL NUMBERS n SUCH THAT EITHER THE DESCRIPTION ON PAGE n IS NOT GENUINE, OR THE DESCRIPTION IS GENUINE AND n DOE NOT BEONG TO THE SET SO DESCRIBED.

Let S be the set of all numbers described by this description. For any n, if the description on Page n is not genuine, then n automatically belongs to the set S. If the description on Page n is genuine, then n belongs to S if and only if n does not belong to the set described on Page n.

Thus the above description specifies for each n whether n belongs to the set S or not, hence the description must be genuine, right? Now supposed this description in on Page 23. What happens then?

Problem 6. How do you get out of that one?

Two Systems of Set Theory

Coming back to the set paradoxes, their proposed resolutions led to two important systems of set theory, the Theory of Types, and the system of Zermelo set theory, later expanded by Fraenkel.

The Theory of Types was expounded in the monumental three-volume *Principia Mathematica* by Bertrand Russell and Alfred North Whitehead. In this theory, one starts with a set of individual elements which are not necessarily sets. These elements are classified as of type 0 (type zero). Any set of elements of type 0 is called a set of type 1. Any set of sets of type 1 is called a set of type 2, and so forth. For any n, the set of all sets of type n is a set of type $n+1$. There is thus no such thing as the set of all sets; rather, for each n, there is the set of all sets of type n, (which of course is of type $n+1$). The Russell paradox cannot even be stated in this system, since in this system there is no such thing as the set of all ordinary sets. [Of course, in this system, every set is ordinary, since any set S is of some type n, hence can be a member only of sets of type $n+1$, hence cannot be a member of itself, for it itself is of type n.]

Ernst Zermelo took a completely different, and far less complicated approach, building on an earlier system of Gottlob Frege, which we must first discuss.

A key axiom of Frege's system was that given any property, there exists the set of all things having that property. This principle is known as the *abstraction principle*, and it is precisely the principle which led to a contradiction, namely the Russell paradox, since it can be applied to the property of sets that are not members of themselves, and thus we get the existence of the set of all ordinary sets, which we have seen leads to a contradiction.

The sad thing is that only after Frege completed his monumental work on set theory did Russell discover and communicate to Frege the inconsistency of his system! Frege was totally crestfallen by this discovery and regarded his whole system as a total failure! In actuality, his pessimism was quite unjustified, since all his other axioms were quite sound and of great importance to mathematical systems of today. It was merely his abstraction principle that needed modification, and this modification was carried out by Zermelo, who replaced the abstraction principle by the following axiom:

Z_0 [*Limited abstraction principle*]. For any property P and any set S, there exists the *set of all elements of* S having property P.

This modification does not appear to lead to any inconsistencies.

Here are some other axioms of Zermelo's system:

Z_1 [*Existence of the empty set*]. There is a set that contains no elements at all.

Before stating other Zermelo axioms, let me tell you a funny incident concerning Z_1. As a graduate student, I had a course in set theory. When the professor introduced the axioms, the first was the existence of the empty set. At that point, brash student that I was, I raised my hand and said, "The empty set must exist, since if it didn't the set of all empty sets would be empty, and we would then have a contradiction." The teacher then explained that without some axioms to the effect, the existence of the set of all empty sets could not be proved – indeed, one could not prove the existence of any set at all! Of course he was right. The interesting thing, though, is that if instead of Z_1, Zermelo had taken the axiom, "There exists a set," the existence of the empty set would follow (by using the limited abstraction principle Z_0).

Problem 7. Why does it follow?

Here is another axiom of Zermelo's system:

Z_2 [*Pairing Axiom*]. For any pair of sets x and y, there is a set that contains both x and y.

Problem 8. Prove that for any sets x and y, there is a set whose elements are just x and y. [This set is denoted $\{x, y\}$.]

Problem 9. Prove that for any set x, there is a set (denoted $\{x\}$) whose only element is x. [Such a set is called a *singleton.*]

Some interesting things follow from just the axioms Z_0, Z_1, and Z_2. For one thing, we can get a model of the natural numbers. For zero, Zermelo took the empty set. For one, he took $\{0\}$, i.e. the set whose only element is the empty set; for two, he took $\{1\}$ (which is $\{\{0\}\}$), for three he took $\{2\}$ (the set $\{\{\{0\}\}\}$), and so forth. Thus for each natural number n, the number $n + 1$ was the singleton $\{n\}$. Thus n is denoted by 0 enclosed in n pairs of brackets.

Later on, John Von Neumann modified Zermelo's construction of the natural numbers, and his is the system used today. Von Neumann defined the natural numbers in such a way that each natural number was the set of all lessor natural numbers. Thus 0 is the empty set; 1 is the set $\{0\}$ (as it was for Zermelo), but 2 is the set $\{0, 1\}$ (the set whose only members are 0 and 1); 3 is the set $\{0, 1, 2\}, \ldots, n + 1$ is the set $\{0, 1, \ldots, n\}$. A technically useful property of Von Neumann's system is that each natural number n consists of exactly n elements (compared to Zermelo's system, in which 0 has no elements, and for each positive n, the number n has just one element, namely the number $n - 1$. More important, however, is the fact that Von Neumann's definition of the natural numbers is easily extended to the definition of the *infinite* ordinals, which play a key role in set theory.

Solutions to the Problems of Chapter 3

1. Suppose I told you that there is a man who is more than six feet tall and also less than six feet tall. How would you explain that?

   The obvious answer is that I must be either mistaken or lying! There obviously couldn't be such a man. Similarly, there couldn't be a barber, given the contradictory information given about him. Thus the answer to the paradox is that there was no such barber.
2. The answer is that the notion of *describable* is not well-defined.
3. The answer is that if that so-called "description" is written on some page, say page 13, then it is not well-defined, since it gives contradictory information about whether 13 belongs to the set or not. Thus it is not a genuine description, rather it is what is called a *pseudo-description.*

   The curious and interesting this is that if those same words are not written on any page of the book, it is a genuine description! But writing it on some page of the book makes it a pseudo-description.
4. If Hypergame were well-defined, we would have a contradiction; hence it is not well-defined. Yes, given a set of games not containing Hypergame, one can well-define a Hypergame *for the set,* but one cannot well define Hypergame for a set that already contains Hypergame.

5. Suppose that in a 1-1 correspondence between A and $\mathcal{P}(A)$, there is an element a of A such that S_a is the set of all normal elements. First we must see that S_a cannot be the empty set $\emptyset$. If it were, then there would only be one element in the path starting from a, namely the path consisting of a itself, a path that terminates immediately. So a would have to be normal. But a cannot be normal, because that would put it into the empty set S_a, which contains no elements. So assuming S_a empty led to a contradiction.

   On the other hand, if S_a were nonempty, a would again have to be normal, since in any path starting with a the next term of the path must be a member a_1 of S_a, and a_1 must be normal (since we are assuming that all elements of S_a are). Thus all paths starting with a_1 will terminate, so that a itself must be normal. Since a is normal, then a must be a member of S_a (since all normal elements of A are in S_a). Hence we can construct the infinite path, $a, a, a, \ldots, a, \ldots$, which makes a not normal! Consequently, no S_a can be the set of all normal elements, which is the same thing as saying that no $a \in A$ can be matched with the subset of A [i.e. element of $\mathcal{P}(A)$] consisting of all the normal elements of A. Hence, since the set of normal elements is a well-defined subset of A, there can be no 1-1 correspondence between A and the full power set of A, $\mathcal{P}(A)$. This proves Cantor's Theorem for an arbitrary set A.
6. The answer is that the very notion of a genuine description is not well-defined!
7. Let P be any property that holds for no set x at all (such as $x \neq x$). If we assume the existence of a set S, then by the limited abstraction principle, there exists the set of all elements of S having this property P, and this set is the empty set.
8. Given the existence of a set S that contains x and y, by the limited abstraction principle, there exists the set of all elements z of S such that $z = x$ or $z = y$. This set is $\{x, y\}$.
9. Given a set S that contains both x and y, there exists the set of all elements z of S such that $z = x$. This set is $\{x\}$. Alternatively, this follows from the last problem, by taking x and y to be the same element.

4
Further Background

Before embarking on the subject of mathematical logic proper, some more basic mathematical topics are in order.

Relations and Functions

We are using the notation $\{x, y\}$ for the set whose element are x and y and no others. The *order* of x and y is immaterial; the set $\{x, y\}$ is the same as the set $\{y, x\}$. In contrast to this, we need the notion of an *ordered* pair (x, y) [note the use of parentheses instead of curly brackets], which consists of two elements x and y, but x is designated as the *first* element of the ordered pair, and y is designated as the *second* element. Now the order is crucial; in general, the ordered pair (x, y) is *distinct* from (y, x) [in fact, they are the same only if x and y are the same element].

The situation is similar with three or more elements. In the set $\{x, y, z\}$, whose elements are x, y, and z, the order doesn't matter, but in the *ordered* triple (x, y, z) [again parentheses instead of curly brackets] the order is crucial; x, y, and z are respectively designated as the first, second and third elements of the ordered triple.

Consider now a *binary* relation R, i.e. a relation between two elements x and y (such as "x loves y," or "x is greater than y," or "$x = y + 1$," or "$x = y^2$"). The relation R determines a unique set of ordered pairs, namely the set of all ordered pairs (x, y) such that x stands in the relation R to y. When we write $R(x, y)$, we will mean that x stands in the relation R to y. In some cases $R(x, y)$ is usually written xRy; for example, in the "less than" relation between numbers, one writes $x < y$ rather than $<(x, y)$. Also, for the identity relation, $=$, one writes $x = y$ rather than $=(x, y)$.

In many modern treatments of set theory, one identifies a binary relation R with the set of all ordered pairs (x, y) that stand in the relation R, and we shall sometimes do that.

For a *trinary* relation R (also called a relation of three arguments, or a relation of degree 3), we write $R(x, y, z)$ to mean that the ordered triple (x, y, z) stands in the relation R. Again, one sometimes identifies a trinary relation R with the set of all ordered triples that stand in the relation. Similar remarks apply to n-ary relations, i.e. relations of n arguments. An ordered triple (x, y, z) is also sometimes called a three-tuple, and $(x_1, x_2, \ldots, x_n)$ is similarly called an n-tuple.

Functions

A function f of one argument is a correspondence through which to every element x of a set S_1 there corresponds a unique element denoted $f(x)$ in a set S_2. In many cases, the sets $S_1$and S_2 are the same. As examples of functions in which S_1 and S_2 are each the set of natural numbers, we could have $f(x) = x + 5$, or $f(x) = x^2$, or $f(x)$ equals the first prime greater than x (in which case $f(8) = 11$, $f(12) = 17$).

One can picture a numerical function f as a computing machine in which one feeds in a number x as input, and out comes a number denoted $f(x)$.

In modern treatments of set theory, one defines a function f of one argument to be a single-valued relation, i.e. a relation such that for each x, there is one and only one y such that (x, y) stands in the relation, and this unique y is denoted $f(x)$.

More generally, for any positive n, a function f of n arguments is a correspondence which assigns to each n-tuple $(x_1, x_2, \ldots, x_n)$ of elements of some set, a unique element denoted $f(x_1, x_2, \ldots, x_n)$ in another set.

For functions of two arguments, one sometimes writes xfy. For example, for the addition function on the natural numbers, one writes $x + y$ rather than $+(x, y)$, and similarly with multiplication and subtraction.

Mathematical Induction

Before explaining this important principle, here is a little story. A certain man was in quest of immortality. He read all the occult books on the subject, but none of them seemed to offer any *practical* solution. Then he heard of a great sage in the East who knew the secret of immortality. After a twelve year search, he finally found him, and asked, "Is it really possible to live forever?" "Quite easily," the sage replied, "providing you do two things." "What are they," the man asked eagerly. "The first thing is to make only true statements in the future. Never make a false one. That's a small price to pay for immortality, isn't it?" "Yes, indeed!" replied the man. "And what is the second thing?" "The second thing," replied the sage, "is to now say 'I will repeat this sentence tomorrow!' If you do those two things, I guarantee you will live forever!"

The man thought about this for a moment, and said, "Of course if I do those two things, I will live forever! If I now truthfully say 'I will repeat this sentence tomorrow,' then I will indeed repeat that sentence tomorrow, and if I am truthful then, then I will repeat the sentence the next day, and so forth. But your solution is not practical! How can I be sure of truthfully saying that I will repeat the sentence tomorrow, since I don't know for sure if I will be alive tomorrow? No, your solution is not practical!"

"Oh," said the sage, "you wanted a *practical* solution! No, I don't deal with practice. I deal only in theory."

Here is a related puzzle. Let us visit a land in which every inhabitant is one of two types, type T or type F. Those of type T make only true statements; everything they say is true. Those of type F make only false statements; everything they say is false. [In many of my puzzle books, those of type T were called *knights*, while those of type F were called *knaves.*] One day one of the inhabitants said, "This is not the first time I have said what I am now saying."

Problem 1. Which type was the inhabitant who said that? Is he of type T or F?

Both the above story and the above puzzle illustrate the principle of mathematical induction. As still another illustration, suppose I tell you that on a certain planet it is raining today, and that on any day on which it rains, it rains the next day as well – in other words, it never rains on any day without raining the next day as well. Doesn't it obviously follow that on the planet it will rain every day after the first day that it rains?

The principle of mathematical induction is that if a property holds for the number 0 and if it never holds for any number n without holding for $n + 1$, then it must hold for all natural numbers.

Still another illustration, with an additional clever and amusing twist: Imagine that we are all immortal and that we live in the good old days in which the milk man would deliver milk to the house and the house wife would leave a note to the milk man telling him what to do. A certain housewife left the following note:

ON ANY DAY THAT YOU LEAVE MILK, BE SURE TO LEAVE MILK THE NEXT DAY AS WELL!

Several days went by, and one day she met the milkman and asked why he didn't obey her order. The milkman replied, "I never disobeyed your order! Your order was to the effect that I should never leave milk on any day without leaving milk the next day as well. Well, I never left milk on any day and didn't leave milk the next day. I never left milk at all, nor did you ever tell me I should!"

The milkman was absolutely right! Her note was insufficient. What she *should* have written is:

(1) ON ANY DAY THAT YOU LEAVE MILK, BE SURE TO LEAVE MILK THE NEXT DAY TOO.
(2) LEAVE MILK TODAY.

That would indeed have guaranteed permanent delivery!

When I told the above story to my friend the computer scientist Dr. Alan Tritter, he came up with the following delightful alternative, which illustrates what might be called the *Turing Machine*, or the *recursive* approach. His note had only one sentence:

LEAVE MILK TODAY AND READ THIS NOTE AGAIN TOMORROW.

Complete Induction

A variant of the principle of mathematical induction is the principle of *complete mathematical induction*, which is this: Suppose a property P of natural numbers is such that for every natural number n, if P holds for all natural numbers less than n, then P holds for n. *Conclusion*: P holds for all natural numbers.

Problem 2. (a) Using the principle of mathematical induction, prove the principle of complete mathematical induction. (b) Conversely, show that the principle of mathematical induction logically follows from the principle of complete mathematical induction.

The Least Number Principle

Until further notice, "number" will mean *natural number*.

The *least number principle* is that every non-empty set of (natural) numbers contains a least number.

This principle is equivalent to the principle of mathematical induction in the sense that either one is a logical consequence of the other.

Problem 3. (a) Taking the principle of mathematical induction as an axiom (which many mathematical systems do), prove the least number principle. (b) Conversely, taking the least number principle as an axiom, prove the principle of mathematical induction.

The Principle of Finite Descent

Suppose a property P is such that for any natural number n, if P holds for n, then P also holds for some natural number less than n. What follows is that P doesn't hold for any number! This is known as the *principle of finite descent*.

Problem 4. Show that the principle of finite decent is equivalent to the principle of mathematical induction.

Limited Mathematical Induction

Problem 5. Suppose that P is a property and n is a number such that the following two conditions hold:

(1) P holds for 0.
(2) For any number x less than n, if P holds for x, then P holds for $x + 1$.

Using mathematical induction, prove that P holds for all numbers less than or equal to n.

Of course there is also the principle of mathematical induction for the *positive integers*, namely that if a property holds for the number 1, and if for every positive integer n, if P holds for n, then P holds for $n + 1$, then, from those two facts it follows that P holds for all positive integers.

A Cute Paradox

Using mathematical induction, I will prove to you that given any finite non-empty set of billiard balls, they must all be of the same color! Here is the proof: For any positive n, let $P(n)$ be the property that for *every* set of n billiard balls, they are all of the same color.

Obviously, given any set containing just one billiard ball, all members of the set are of the same color (since all of them are the same single element), hence $P(1)$ is true, or we might say "P holds for the number 1." Next we must show that for any n, if P holds for n, then P also holds for $n + 1$. So suppose n is a number for which P holds. Now consider any set of $n + 1$ billiard balls. Number them from 1 to $n + 1$.

$$\begin{array}{cccccc} \circ & \circ & \ldots & \circ & \circ \\ 1 & 2 & & n & n+1 \end{array}$$

The balls 1 through n all have the same color (by the assumption that P holds for n, i.e. for *any* set of n balls, they are of the same color). Likewise the n balls numbered from 2 to $n + 1$ are of the same color. Thus ball $n + 1$ is of the same color as ball 2, which is the same color as the rest of the balls from 1 to n. Thus all $n + 1$ balls are of the same color, which proves that the property P also holds for $n + 1$. Thus we have shown $P(1)$, and that for all n, $P(n)$ implies $P(n + 1)$; hence, by mathematical induction, P holds for every positive number n!

Problem 6. What is the error in the above proof?

Exercise 1. Suppose Q is a property of *sets* of (natural) numbers, which satisfies the following two conditions:

(1) Q holds for the empty set.
(2) For any finite set A and any number n not in A, if Q holds for A, then Q holds for $A \cup \{n\}$ (which we recall is the set of all elements of A together with the element n).

Prove that Q holds for all finite sets of natural numbers. [Hint: Define $P(n)$ to mean that every set of n numbers has property Q. Then use mathematical induction to show that $P(x)$ holds for every natural number x.]

Exercise 2. We earlier stated without proof that no finite set can be put into a 1-1 correspondence with any of its proper subsets. Prove this by mathematical induction on the number of elements of the set.

The next two problems are not necessary for the rest of this book but are (hopefully) of independent interest.

Problem 7. [An induction within an induction] Consider a relation $R(x, y)$ between (natural) numbers such that:

(1) $R(n, 0)$ and $R(0, n)$ holds for every number n.
(2) For all numbers x and y, if $R(x, y+1)$ and $R(x+1, y)$, then $R(x+1, y+1)$.

Prove that $R(x, y)$ holds for every x and y. [Hint: Call a number y *special* if $R(x, y)$ holds for every x. Prove by induction that every y is special. (In going from y to $y+1$, another induction is involved!)]

Problem 8. [Another double induction principle] This time we are given a relation $R(x, y)$ satisfying the following two conditions:

(1) $R(x, 0)$ holds for every x.
(2) For all x and y, if $R(x, y)$ and $R(y, x)$, then $R(x, y+1)$.

Prove that $R(x, y)$ holds for all x and y. [Hint: Call a number x *left normal* if $R(x, y)$ holds for every y, and *right normal* if $R(y, x)$ holds for every y, First show that every right normal number is also left normal. Then show by induction that every number is right normal (hence also left normal).]

A Ball Game

Imagine that you live in a universe in which everyone is immortal, except for the possibility of being executed. You are required to play the following game: There is an infinite supply of pool balls available, each bearing a positive integer. For each positive n, there are infinitely many balls numbered n. In a certain box there are finitely many of these balls, but the box is expandable and has infinite capacity. Each day you are required to throw a ball out of the box (never to be put back into the box again) and replace it by an *finite* number of *lower-numbered* balls; for example, you may throw out a 47 and replace it by a million 46's. If the box ever gets empty, you get executed!

Problem 9. Is there a strategy by which you never get executed, or is execution inevitable sooner or later?

König's Lemma

The mathematician Dénes König stated and proved an interesting result about what are called *trees*, which have significant applications to mathematical logic, as we will see.

A tree consists of an element a_0 called the *origin*, which is connected to finitely many (possibly zero) or denumerably many elements called the *successors* of a_0, each of which in turn has finitely or denumerably many successors, and so forth. We say that an element x is a predecessor of y if y is a successor of x. The origin has no predecessor, and every other element of the tree has one and only one predecessor. The elements of the tree are called *points*. An element with no successor is called an *end point*, and a *junction point* otherwise. The origin of the tree is said to be of level 0, its successors are of level 1, the successors of these successors are of level 2, and so forth. Thus for each number n, the successors of a point of level n are of level $n + 1$. Trees are displayed with the origin at the top and the tree grows downward. By a *path* of the tree is meant a finite or denumerable sequence of points, beginning with the origin, and such that each of the other terms of the sequence is a successor of the preceding term. Since a point cannot have more than one predecessor, it follows that for any point x, there is one and only one path from x up to the origin a_0 (which implies that there is also only one path down from the origin a_0 to the point x). By the *length* of a finite path is meant the level of its endpoint. By the *descendants* of a point x are meant the successors of x, together with the successors of the successors, together with the successors of these, and so forth. Thus y is a descendant of x if and only if there is a path to y that goes through x.

Problem 10. Suppose that we are given that for any positive integer n, there is at least one path of length n (and thus every level is hit by at least one path). Does it necessarily follow that there must be at least one infinite path?

Finitely Generated Trees

Let us note that if a point x of a tree has only finitely many successors $x_1, \ldots, x_k$, then if each successor of x has only finitely many descendants, then x has only finitely many descendants, because if n_1 is the number of descendants of x_1, n_2 is the number of descendants of $x_2, \ldots$, and n_k is the number of descendants of x_k, then the number of descendants of x is $k + n_1 + \cdots + n_k$, a finite number.

A tree is said to be *finitely generated* if each point has only finitely many successors (though it might have infinitely many descendants).

Problem 11. Suppose a tree is finitely generated.

(a) Does it necessarily follow that each level contains only finitely many points?
(b) Which, if either, of the following two statements implies the other?
  (1) For each n there is at least one path of length n.
  (2) The tree has infinitely many points.

Now, in the solution to Problem 9, we had an example of a tree such that for each positive n, there is a path of length n, yet there was no infinite path. Of course that tree was not finitely generated, since the origin had infinitely many successors. For finitely generated trees, the situation is quite different.

König's Lemma. A finitely generated tree with infinitely many points must have an infinite path.

Problem 12. Prove König's lemma. [Hint: Recall that if a point x has only finitely many successors, and if each of the successors has only finitely many descendants, then x has only finitely many descendants.]

A tree is called *finite* (not to be confused with *finitely generated*) if it has only finitely many points.

The following result of the Dutch mathematician L. E. J. Brouwer [1927] is closely related to König's lemma.

The Fan Theorem. If a tree is finitely generated and if all paths are finite, then the whole tree is finite.

Problem 13. Prove the Fan Theorem.

Discussion. Brouwer's Fan Theorem is in fact the *contrapositive* of the König lemma. [By the *contrapositive* of an implication "p implies q" is the proposition "not q implies not p."] Now, the logic we are using in this book is known as *classical* logic, a fixture of which is that any implication is equivalent to its contrapositive. There is a weaker system of logic known as *intuitionist logic*, of which Brouwer was a major representative, in which not all implications can be shown to be equivalent to their contrapositives. In particular, the Fan Theorem can be proved in intuitionist logic, but König's Lemma cannot. Another proof of the Fan Theorem will be given shortly, one which is intuitionistically acceptable. Unfortunately, we will not be able to say more about intuitionist logic in this work.

Here is a cute application of the Fan Theorem:

Problem 14. [Ball Game Revisited] The Fan Theorem provides a particularly elegant alternative solution to the Ball Game problem, one which at least in my opinion is much neater than the solution already given, which used mathematical induction. Can the reader find it?

Generalized Induction

The principle of complete mathematical induction, which is a result about the natural numbers, has an important generalization to arbitrary sets, even non-denumerable ones.

We now consider an arbitrary set A of any size and a relation $C(x, y)$ between elements of A, which we read, "x is a *component* of y." [The notion of *component* has many applications to logic, set theory and number theory. In set theory the components of a set are its elements. In some applications of number theory, the components of a natural number n are the numbers less than n. In others, a number x is a component of y if $x + 1 = y$. In mathematical logic, the components of a formula are its so-called sub-formulas.]

Given a component relation $C(x, y)$ on a set A, by a *descending chain* is meant a finite or denumerable sequence of elements of A such that each term of the sequence, other than the first, is a component of the preceding term. Thus for any finite sequence $(x_1, x_2 \ldots, x_n)$ or infinite sequence $(x_1, x_2 \ldots, x_n, \ldots)$, x_2 is a component of x_1, x_3 is a component of x_2, etc.

Generalized Induction Principle

Given a component relation $C(x, y)$ on a set A, a property P of elements of A is said to be *inductive* if for every element x of A, if P holds for all components of x, then P holds for x. It is to be understood that if x has no components at all then P automatically holds for x, since P does hold for all components of x, of which there are none. [We recall that anything said about all elements of the empty set is true!]

We now say that the component relation $C(x, y)$ obeys the *generalized induction principle* if every inductive property P holds for all elements of A, in other words, if for any property P, if P is such that for each element x of A, whenever P holds for all components of x, then P holds for x, then P must hold for all elements of A.

Thus if $C(x, y)$ obeys the generalized induction principle, then for any property P, to show that P holds for all elements of A, it suffices to show that for all elements x of A, if P holds for all components of x, then P holds for x as well.

Let us note that the principle of mathematical induction for the natural numbers is that the component relation $x + 1 = y$ obeys the generalized induction principle.

The principle of complete induction for the natural numbers is that the component relation $x < y$ (x is less than y) obeys the generalized induction principle.

The following result is basic:

Generalized Induction Theorem. A sufficient condition for a component relation $C(x, y)$ on a set A to obey the generalized induction principle is that all descending chains be finite.

Thus if all descending chains are finite, then to show that a given property P holds for all elements of A, it suffices to show that for every element x of A, if P holds for all components of x, then P holds for x as well.

Problem 15. Prove the above theorem. [Hint: Note that if a property P is inductive, then for any element x, if P fails for x, then P must fail for at least one component of x.]

Tree Induction

The Generalized Induction Theorem has an obvious application to trees, even to those with infinitely many points. If all branches of the tree are finite, then to show that a given property P holds for all points of the tree, it suffices to show that for any point x, P holds for x provided P holds for all successors of x. This is a special case of the Generalized Induction Theorem by taking the components of x to be the successors of x.

Problem 16. The generalized induction theorem provides an alternative proof of the Fan Theorem (hence also of König's Lemma). How?

Problem 17. The converse of the Generalized Induction Theorem also holds, that is, if the component relation obeys the generalized induction principle, then all descending chains must be finite. Prove this.

Well-Founded Relations

We have seen that the generalized induction principle is equivalent to the condition that there be no infinite descending chains (in Problems 15 and 17). Each of these two is in turn equivalent to the interesting third condition that we will now consider.

Again we consider a relation $C(x, y)$ on a set A. For any subset S of A, an element x of S is called an *initial* element of S (with respect to the relation $C(x, y)$ understood) if x has no components in S. The relation $C(x, y)$ is called *well-founded* if every non-empty subset of A has an initial element.

And now for a lovely result:

Theorem. For any component relation $C(x, y)$ on a set A, the following three conditions are all equivalent:

(1) The generalized induction principle holds.
(2) All descending chains are finite.
(3) The relation is well-founded.

Problem 18. Prove the above.

Compactness

We now turn to another principle that is quite useful in mathematical logic and set theory.

Consider a universe V in which there are denumerably many people. The people have formed various clubs. A club C is called *maximal* if it is not a *proper* subset of any other club. Thus if C is a maximal club, if we add one or more people to the set C the resulting set is not a club.

Problem 19. (a) Assuming there it at least one club, is there necessarily a maximal club? (b) Suppose that the number of people in V is finite instead of denumerable. Would that change the answer to (a)?

Problem 20. Again we consider the case that V contains denumerably many inhabitants, and we assume that we are given the additional information that a set S of people is a club if and only if every finite subset of S is a club. Then it does follow that if there is at least one club, there is a maximal club; better yet, every club is a subset of a maximal club. The problem is to prove this. Here are the key steps:

(1) Show that under the given conditions, every subset of a club is a club.
(2) Since V is denumerable, its inhabitants can be arranged in a denumerable sequence $x_0, x_1, \ldots, x_n, \ldots$. Now, given any club C that exists, define the following infinite sequence $C_0, C_1, C_2, \ldots, C_n, \ldots$ of clubs.
   (a) Take C_0 to be C.
   (b) If adjoining x_0 to C_0 yields a club, take C_1 to be this club, $C_0 \cup \{x_0\}$; otherwise let C_1 be C_0 itself. Then consider x_1. If $C_1 \cup \{x_1\}$ is a club, let C_2 be this club, otherwise let C_2 be C_1. Continue thus way through the series, that is, once C_n has been defined, take C_{n+1} to be $C_n \cup \{x_n\}$ if that set is a club, otherwise take C_{n+1} to be C_n. It is obvious that each C_n is a club.
   (c) Now take C^* to be the union of C with all of the people who were added in to any C_i of the sequence at some stage, which is the same as

the set of all people who belong to at least one of the infinitely many clubs $C_0, C_1, C_2, \ldots, C_n, \ldots$. Then prove that C^* is a club, and in fact a maximal club.

Consider now an arbitrary denumerable set A and a property P of subsets of A. The property P is called *compact* if it is the case that for every subset S of A, P holds for S if and only if P holds for all finite subsets of S. Also, a collection Σ of subsets of A is called *compact* if the property of being a member of Σ is compact, in other words, if for any subset S of A, S is in Σ if and only if all finite subsets of S are in Σ. By a *maximal* element of Σ is meant a member of Σ that is not a *proper* subset of any element of Σ.

In the last problem, we were given a denumerable universe V and a collection of subsets called *clubs*, and we were given that a subset S of V is a club if and only if all finite subsets of S are clubs, in other words, we were given that the property of being a club is *compact*, from which we could deduce that any club is a subset of a maximal club. Now, there is nothing special about *clubs* that makes our argument go through. The same reasoning yields the following:

Denumerable Compactness Theorem. For any compact property P of subsets of a denumerable set A, any subset S of A having property P is a subset of a maximal set that has property P.

Note: The above result actually holds even if A is a non-denumerable set (whether finite or infinite), but the more advanced result for infinite non-denumerable sets is not needed for anything in this book. The reader should also be able to see that the proof just given can be modified slightly to work when V is finite.

Discussion

In mathematical logic, we deal with a denumerable set of symbolic sentences, some subsets of which are deemed inconsistent. Now, since the mathematical logic we will be studying in this book considers only proofs employing a finite number of sentences, any proof that a given set of sentences is inconsistent uses only finitely many members of the set of those sentences. Thus a denumerable set S is defined to be *consistent* if and only if all finite subsets of S are consistent. In other words, consistency is a compact property, a fact that will prove very important later on!

Problem 21. For any property P of sets, and for any set S, define $P^*(S)$ to mean that all finite subsets of S have property P. Prove that regardless of whether P is compact or not, the property P^* is compact.

Solutions to the Problems of Chapter 4

1. If he were of type T, then he really would have said that sometime before, as he said, and when he said that before, he was still of type T, hence he had said it sometime before that, hence sometime before even that, and so forth. Thus unless he lived infinitely far back into the past, he cannot be of type T. Thus he is of type F.
2. (a) Given the principle of mathematical induction, we are to prove the principle of complete mathematical induction.

   We consider a property P such that for every number n, if P holds for all numbers less than n, then P also holds for n. We are to show that P holds for all numbers.

   Now, there are no natural numbers less than 0, so that the set of natural numbers less than zero is empty. And we already know that any property holds for all elements of the empty set, so that P holds for all natural numbers less than zero (since there aren't any). Thus, by our assumption of the induction premise of complete mathematical induction (i.e. that, if P holds for all numbers less than n, then P also holds for n), P holds for zero. Now it is not easy to directly show that, for all n, if P holds for n, then P holds for $n+1$, and so we resort to a useful trick. We consider a property Q stronger than P (in the sense that any number having property Q also has property P), and we use mathematical induction on the property Q, i.e. we show that $Q(0)$ (0 has the property Q) and that $Q(n)$ implies $Q(n+1)$ for every n. In fact we define $Q(n)$ to mean that P holds for n and all numbers less than n. Of course $Q(n)$ implies $P(n)$.

   Q vacuously holds for all natural numbers less than zero, since there aren't any, and P holds for 0, as we have seen; thus Q holds for 0. Now suppose n is a number for which Q holds. This means that P holds for n as well as for all numbers less than n; therefore P holds for all numbers less than $n+1$, so that by the induction hypothese of complete induction, P holds for $n+1$. Thus P holds for $n+1$, and all numbers less than $n+1$, which means that Q holds for $n+1$. This proves that if Q holds for n, it also holds for $n+1$, and so, by mathematical induction, Q holds for all natural numbers, which of course implies that P holds for all natural numbers.

   (b) Conversely, assume the principle of complete mathematical induction. We are then to infer the principle of mathematical induction. Well, suppose that P is a property such that:

   (1) $P(0)$,

   (2) $P(n)$ implies $P(n+1)$, for all n.

   We are to show that P holds for every n. To do this, it suffices to show that for any number n, if P holds for all numbers less than n. it holds for n (and then by the assumed principle of complete mathematical induction, P will hold for every n).

Well, suppose P holds for all numbers less than n. For the case of $n = 0$, we are already given that P holds for 0. And so suppose $n \neq 0$. Then $n = m + 1$ for the number $m = n - 1$. Since P holds for all numbers less than n, then P holds for m. Hence, by (2), P holds for $m + 1$, which is n. This proves that if P holds for all numbers less than n, then P holds for n. Then by the assumed principle of complete mathematical induction, P holds for all numbers.

3. (a) Assume the principle of mathematical induction. We are to then prove the least number principle.

Define $P(n)$ to mean that any set A that contains a number x no greater than n must have a least element. (Note: To say that A contains a number x no greater than n just means that A contains some number x that is less than or equal to n.) First we show by mathematical induction that every number n has property P.

Suppose A is a set that contains a number x that is no greater than zero. Then x must be zero; hence 0 is an element of A, and thus the least element of A. This proves that P hold for 0.

Next, suppose that P holds for n. Now let A be any set that contains a number no greater than $n + 1$. We must show that A contains a least element.

Either A contains an element less than $n + 1$ or it doesn't. If it doesn't, then $n + 1$ is the only element of A, hence is the least element of A. On the other hand, if it does, then it contains some number not greater than n, hence by our assumption of $P(n)$, A must contain a least element. This proves that $P(n)$ implies $P(n + 1)$, and, since $P(0)$ holds, then, by mathematical induction, it must be true that $P(n)$ holds for every n.

Now let A be any non-empty set of numbers. Then A contains *some* number n. Hence A contains a number no greater than n (namely n itself). Thus, since P holds for n (as we have seen), A must contain a least element. The proves the least number principle.

(b) Conversely, suppose we start out with the least number principle. We wish to then prove the principle of mathematical induction.

We are given a property P such that:

(1) $P(0)$,

(2) $P(n)$ implies $P(n + 1)$, for every n.

We are to show that P holds for every n.

If P fails to hold for every n, then there is at least one number for which P fails. Hence, by the least number principle (which we are assuming), there must be a *least* number n for which P fails. This number n cannot be zero, hence $n = m + 1$ for the number $m = n - 1$. Since m is less than n and n is the least number for which P fails, then P cannot fail for m, and so P must hold for m. Thus P holds for m but fails for $m + 1$ (which is n), contrary to (2) above ($P(n)$ implies $P(n + 1)$). This is a contradiction, and so P cannot fail for any number. Thus P holds for every number.

4. The principle of finite descent really says the same thing as the principle of complete mathematical induction, because the statement "If P fails for n, then P fails for some number less than n" is logically equivalent to "If P holds for all numbers less than n, then P holds for n." Thus the principle of finite descent is but another way of stating the principle of complete mathematical induction, which in turn is equivalent to the principle of mathematical induction, as we have seen.
5. We use the standard abbreviations "$x < y$" for "x is less than y," and "$x \leq y$" for "is less than or equal to y."

   We are given a number n and a property P satisfying the following two conditions:

   (1) $P(0)$,

   (2) For all numbers x, if $x < n$ and $P(x)$, then $P(x+1)$.

   We are to show that P holds for all x less than or equal to n.

   Let $Q(x)$ be the property that if $x \leq n$, then $P(x)$. We will use mathematical induction on Q to show that Q holds for all x. Since $P(0)$ is true, then of course $0 \leq n$ implies $P(0)$, and thus $Q(0)$ holds.

   Next, suppose that $Q(x)$ holds. We are to show that $Q(x+1)$ must also hold. Well, suppose $x+1 \leq n$. We are to show $P(x+1)$. Since $x+1 \leq n$, then $x < n$, so that of course also $x \leq n$. Since $x \leq n$, and $x \leq n$ implies $P(x)$ (which is the induction assumption $Q(x)$), then $P(x)$ holds. Since $x < n$ and $P(x)$ holds, then $P(x+1)$ holds (by the given condition (2)). This proves that if $x+1 \leq n$, then $P(x+1)$ holds, which is the property $Q(x+1)$. Thus $Q(x)$ implies $Q(x+1)$, and since $Q(0)$ also holds, then, by mathematical induction, $Q(x)$ holds for all x. In particular $Q(x)$ holds for all $x \leq n$, which means that $P(x)$ holds for all $x \leq n$.
6. We are numbering the billiard balls $B_1, \ldots, B_n, B_{n+1}$. The proof fails because the induction fails to go from 1 to 2. Of course, for any n greater than 1, the sets $\{B_1, \ldots, B_n\}$ and $\{B_2, \ldots, B_n, B_{n+1}\}$ overlap (they have the elements $B_2, \ldots, B_n$ in common), but this is not so for $n = 1$. The sets $\{B_1\}$ and $\{B_2\}$ do not overlap. Thus it is not true that for all $x \leq n$, $P(x)$ implies $P(x+1)$. It is not true for $n = 1$, since it is not true that $P(1)$ implies $P(2)$. $P(1)$ is true, but $P(2)$ is false, since it is not true that any two billiard balls are of the same color (indeed, if it were true, then all billiard balls would indeed be of the same color!).
7. We are to prove by induction that every number y is special. Well, we are given that $R(x, 0)$ holds for every x; hence 0 is special. Now suppose that y is special. We are to show that $y+1$ is special. Thus we must show that $R(x, y+1)$ holds for every x. We show this by induction on x (which is where an induction within an induction comes in!).

   Since $R(0, n)$ holds for every n, then $R(0, y+1)$ holds. Now suppose that x is such that $R(x, y+1)$. Also, $R(x+1, y)$ holds, under the induction assumption that y is special. Thus $R(x, y+1)$ and $R(x+1, y)$ both hold, so that by the given condition (2), we see that $R(x+1, y+1)$. This proves that $R(x, y+1)$ implies $R(x+1, y+1)$. And so by induction $R(x, y+1)$ holds for all x, which means that $y+1$ is special. This proves that if y is special, so is $y+1$, which completes the inductive prove that every number y is special. Thus $R(x, y)$ holds for all x and y.

8. We first show that every right normal number x is also left normal. Well, suppose x is right normal. We show by induction on y that $R(x, y)$ holds for all y.

   We are given outright that $R(x, 0)$ holds. Now suppose that y is such that $R(x, y)$ holds. Since x is assumed right normal, then $R(y, x)$ holds. Since $R(x, y)$ and $R(y, x)$, then also $R(x, y+1)$. And since $R(x, 0)$ holds, then by induction $R(x, y)$ holds for all y, which means that x is left normal.

   Now we show by mathematical induction that every number y is right normal (hence also left normal).

   By (1), $R(x, 0)$ holds for all x, which tells us that 0 is right normal. Now suppose that y is right normal. Then it is also left normal (as we have seen); hence for every number x, we have both $R(y, x)$ and $R(x, y)$, hence also $R(x, y+1)$, which means that $y+1$ is right normal (since $R(x, y+1)$ holds for every x). This proves that if y is right normal, so is $y+1$, which completes the inductive proof that every number y is right normal. Thus, since every number x is both left and right normal, $R(x, y)$ holds for every x and y.

9. The answer is that the box must become empty sooner or later. One proof of this, which we now give, is by mathematical induction. Another proof will be given later on.

   Call a positive integer n a *losing number* if it is the case that every ball game in which every ball originally in the box is numbered n or less (i.e. every ball game starting with such a box) must terminate. We will now show by mathematical induction that every positive integer n is a losing number.

   Obviously 1 is a losing number (if only 1's are initially in the box, the game must clearly terminate, since any 1 thrown out of the box cannot be replaced by anything, and there can only be finitely many 1's in the box). Now suppose that n is a losing number. We are to show that $n+1$ must then also be a losing number.

   Consider any box in which every ball is numbered $n+1$ or less. One cannot keep throwing out balls numbered n or less forever, since n is assumed to be a losing number. Hence sooner or later one must throw out a ball numbered $n+1$ (assuming there is at least one in the box). Then sooner or later one must throw out another ball numbered $n+1$ (if there are any left). And so, continuing this way, one must sooner or later get rid of all the balls numbered $n+1$, and thus be left with a box in which the highest numbered ball is n or less. And from then on, by the induction assumption that n is a losing number, the process must eventually terminate. This completes the proof that every positive integer

n is a losing number, which means that for any given box, no matter what daily choices are made, the process must eventually terminate.

As previously remarked, another proof will be given later, one which I believe is far neater and more elegant than the one just given.

Remarks. The curious thing about this ball game is that there is no finite limit to how long the player can live (assuming that at least one initial ball is numbered 2 or higher), yet the player cannot live forever. Given any positive integer n, he can be sure to live n or more days, but he still can't live forever.

10. It is interesting that many people give the wrong answer to this question. They believe that there must have to be an infinite path, whereas in fact, there does not have to be one, as the following tree illustrates:

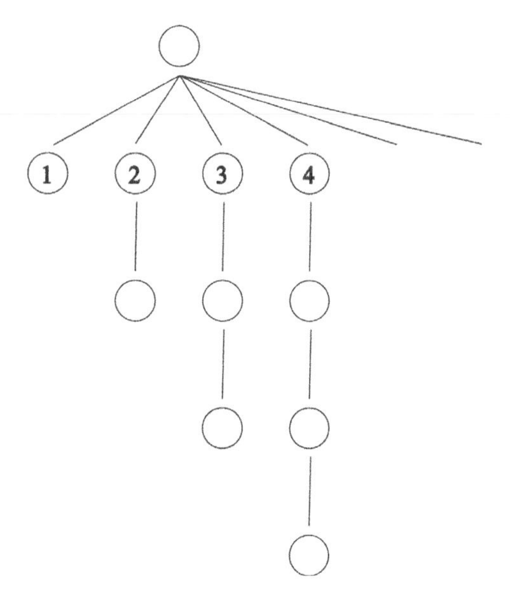

The origin of this tree has denumerably many successors numbered $1, 2, \ldots, n, \ldots$. Every other point of the tree has only one successor. The path through point number 1 stops at level 1. The path through point 2 goes down only two levels, and so forth. That is, for each n, the path through point n stops at level n. Thus, for each n, there is a path of length n, yet none of the paths are infinite.

The situation can be likened to the natural numbers. Obviously for each natural number n, there is a number equal to or greater than n, yet no natural number is itself infinite.

11. (a) Yes, for each n, the nth level contains only finitely many points, as is easily shown by mathematical induction on n: For $n = 0$, there is only one point at level 0, the origin. Now suppose n is such that there are only finitely many points $x_1, \ldots, x_k$ of level n. Let n_1 be the number of successors of $x_1, \ldots, n_k$

be the number of successors of x_k. Then the number of points of level $n+1$ is $n_1 + \cdots + n_k$, which is a finite number. Thus, by mathematical induction, every level contains only finitely many points.

(b) The two statements are equivalent. It is obvious that statement (1) implies statement (2), since statement (1) implies that every level n has at least one point. In the other direction, suppose statement (2) holds, i.e. that there are infinitely many points in the tree. If there were some n such that no path was of length n, then only finitely many levels would have any points at all (since if there is no path of length n, there can be no points at level n or higher). And by (a) each of these levels would have only finitely many points. Thus there would be only finitely many points on the tree, contrary to the assumed condition that there are infinitely many points on the tree.

12. Call a point *rich* if it has infinitely many descendants, and *poor* if it does not. We recall that if a point x has only finitely many successors, if each of those successors is poor, then x must be poor. Thus if x is rich and has only finitely many successors, then it cannot be that all its successors are poor. At least one of them must be rich. Thus in a finitely generated tree, each rich point must have a rich successor. [A rich point with infinitely many successors doesn't necessarily have to have at least one rich successor. For example, in the tree of the solution of Problem 10, the origin is rich, but all its successors are poor.]

    Now, we are considering a tree with infinitely many points and such that each point has only finitely many successors. Since there are infinitely many points, and all points other than the origin are successors of the origin, the origin must be rich, hence must have at least one rich successor x_1 (as we have seen), which in turn has at least one rich successor x_2, which in turn has at least one rich successor x_3, and so forth. We thus generate an infinite path.

13. This is immediate from König's lemma: Suppose a tree is finitely generated and that all paths are finite. If the tree were infinite, then by König's lemma, there would be an infinite path, contrary to the given condition that all paths are finite. Therefore the tree cannot be infinite. It must be finite.

14. To each ball game we associate a tree as follows: We take the origin to be a ball of any higher number than any of the balls initially in the box. We take the successors of the origin to be the balls that are initially in the box. For any ball x that is ever in the box, we take its successors to be the balls that replace it (if any ever do). Since every ball that is ever replaced is replaced by only finitely many balls, the tree is finitely generated. Since all replacements of a ball have lower numbers, each path must be finite. Then by the Fan Theorem, the tree is finite. Hence only finitely many balls ever find themselves in the box (if only for a time). Therefore the game must terminate, for if it didn't, infinitely many balls would be in the box (not necessarily simultaneously) at one point or another, for instance each ball that is thrown out of the box (which would be

an infinite number of balls, if the game went on forever). And that would make the tree we constructed infinite, since it contains all balls that were ever in the box. (Remember we assumed that once a ball was thrown out of the box, that particular ball is never put back into the box, so that each ball thrown out must be different from every other ball thrown out.)

Voila!

15. If an inductive property P fails for x, then it must fail for some component x_1 of x, hence it must also fail for some component x_2 of x_1, and so forth; which would mean that there is an infinite descending chain. Thus if all descending chains are finite, an inductive property cannot fail for any element, and thus holds for all elements of A.
16. We are considering a tree which is finitely generated and is such that all its paths are finite. We must show that the tree itself must be finite.

    For any point x of the tree, we define its components to be its successors. Let $P(x)$ be the property that x is poor (has only finitely many descendants). For any x, since x has only finitely many successors, then if all components (successors) of x are poor, so is x (as we already noted), which means that the property P is inductive! The descending chains of this component relation are the paths of the tree, and we are given that they are all finite. Hence, by the Generalized Induction Theorem, P holds for all points of the tree. In particular, P holds for the origin, so that the origin is poor. Hence there are only finitely many points on the tree.
17. It is obvious that if no component of x begins an infinite descending chain, neither does x (since any chain starting with x has to first go through a component of x). Thus the property of not beginning any infinite chain is inductive. Thus if the generalized induction principle holds, then all elements have this property, which means that no element x begins an infinite descending chain, and so all descending chains are finite.
18. We will show that well-foundedness is equivalent to the condition that there are no infinite descending chains
    (a) In one direction, suppose a component relation $C(x, y)$ on a set A is well-founded. If there were an infinite descending chain, then the subset of A composed of the elements of the chain would have no initial element, contrary to the given condition that the relation C is well-founded.
    (b) In the other direction, suppose that all descending chains are finite. Given any non-empty subset S of A, let x_1 be any element of S. If x_1 has no components in S, we are done; otherwise, x_1 has a component x_2 in S. If x_2 is an initial element of S, we are done; otherwise, x_2 has a component x_3 in S, and so forth. Since there are no infinite descending chains, we must sooner or later come to an initial element x_n of S. Thus the component relation is well-founded.

19. Solutions to the two parts:
    (a) The answer is *no* – there doesn't necessarily have to be a maximal club. For example, it could be that all and only the finite sets are the clubs, and there is obviously no maximal finite set.
    (b) We assume V is finite and that there is a club C. If C is maximal, we are done. Otherwise C is a *proper* subset of a club C_1. If C_1 is maximal, we are done. If not, then C_1 is a *proper* subset of some club C_2, and so forth. This sequence of clubs cannot go on forever, since V is finite. Thus it must end in some maximal club C_n.
20. We are now given that a set is a club if and only if all its finite subsets are clubs.
    (1) First we are to show that every subset of a club C must be a club. Well, consider any subset S of C. Obviously, all subsets of S are also subsets of C. Now let F be any finite subset of S. Since F is clearly also a finite subset of C, it must be a club. Thus all finite subsets of S are clubs; hence S is a club.
    (2) Now we must show that the C^* defined in (b) and (c) just after the statement of this problem in the text of this chapter is a maximal club, indeed a maximal club of which the (arbitrary) club C is a subset. It is immediate that for any number n, the club C_n is a subset of C_{n+1}. Hence for all numbers n and m, if $n < m$, then C_n is a subset of C_m. From this it follows that for any finite set Σ of the clubs $C_0, C_1, C_2, \ldots, C_n, \ldots$, one of them includes all the others, namely the club C_i in the set Σ that has the largest subscript.

       We now consider the set C^* of all people who belong to any of the clubs $C_0, C_1, C_2, \ldots, C_n, \ldots$, and we are to show that C^* is a maximal club. To show that C^* is a club, it suffices to show that every finite subset of C^* is a club. So let F be any finite subset of C^*. Each member of F belongs to one of the clubs $C_0, C_1, C_2, \ldots, C_n, \ldots$, so if F has k elements, we can choose k clubs from our sequence, the first club chosen to contain the 1st element of F, the second club chosen to contain the second element of F, and so forth. Of the finitely many clubs just chosen from the sequence, at least one of them includes all the others (namely any one for which no other club chosen has a larger subscript). Hence that club alone contains all members of F, making F a subset of a club C_n for some n. And since every subset of a club is a club (as shown in (1)), F must be a club. Thus every finite subset of C^* is a club, and therefore C^* is a club.

       As to maximality, we first show that for any person x_n, if $C^* \cup \{x_n\}$ (which we recall is the set of all members of C^* along with x_n) is a club, then x_n must already be a member of C^*. Well, suppose $C^* \cup \{x_n\}$ is a club. Since C_n is a subset of C^*, $C_n \cup \{x_n\}$ is a subset of $C^* \cup \{x_n\}$, and since $C^* \cup \{x_n\}$ is a club (by assumption), then its subset $C_n \cup \{x_n\}$ is a club (as shown in (1)). Since $C_n \cup \{x_n\}$ is a club, then $C_n \cup \{x_n\} = C_{n+1}$. Since $x_n \in C_{n+1}$ and C_{n+1} is a subset of C^*, then $x_n \in C^*$, which is what was to be proved. Thus, for any person x, if $C^* \cup \{x\}$ is a club, x must already be a member of C^*. Thus for any individual x outside the set C^*, the set $C^* \cup \{x\}$ is not a club.

Now consider any set A of individuals such that C^* is a *proper* subset of A. Then A contains at least one individual x who is not a member of C^*; hence $C^* \cup \{x\}$ is not a club. Hence A cannot be a club (for if it were, then its subset $C^* \cup \{x\}$ would be a club, which it isn't). This proves that for any set A of individuals, if C^* is a proper subset of A, then A is not a club. Thus C^* is a maximal club.

21. We are to show that the property P^* is compact, i.e. that a set S has property P^* if and only if all finite subsets of S have property P^*.
    (1) Suppose S has property P^*. Thus all finite subsets of S have property P, and thus also have property P^*. Actually, not only do all finite subsets of S have property P^*, but *all* subsets of S have property P^*. To see this, let A be any subset of S. Then all finite subsets of A are also finite subsets of S, so that all have property P, which means that A has property P^*. Thus all subsets of S have property P^*.
    (2) To show that if all finite subsets of S have property P^*, so does S, we first note that any finite set A having property P^* also has property P, because having property P^*, all finite subsets of A have property P, hence so does A, being a finite subset of A. Thus any finite set that has property P^* also has property P. We have now seen that when all finite subsets of S have property P^*, they all also have property P. But this just means that S has property P^*.

Part II
Propositional Logic

5

Beginning Propositional Logic

The so-called *logistic program*, which is the one we are following, is one that is designed to develop *all* mathematics from just a few principles of logic. The starting point is Propositional Logic, which is the basis of First-Order Logic, as well as of higher order logics.

Propositions can be combined to form more complex propositions by using the so-called *logical connectives*. The principle logical connectives are the following:

(1) $\sim$ negation, not
(2) $\wedge$ conjunction, and
(3) $\vee$ disjunction, either-or
(4) $\supset$ implication, if-then
(5) $\equiv$ equivalence, if and only if

Negation

For any proposition p, by $\sim p$ is meant that p is not true. The proposition $\sim p$ is called the *negation* of p, and is true if and only if p is false, and is false if and only if p is true. These two facts are summarized in the following table, which is called the *truth table* for negation. Here, as in all subsequent truth tables, the letter "T" stands for truth and "F" for falsity.

p	$\sim p$
T	F
F	T

Each row in a truth table corresponds to a particular distribution of truth values for the variables that occur in the formula that the truth table is for. In the case of the truth table for the formula $\sim p$, there is only one variable, which can have only two possible truth values. The first row of this truth table for negation (i.e., the row under the column labels) says that if p is true, then $\sim p$ is false. The second row of the truth table says that if p is false, then $\sim p$ is true. The proposition $\sim p$ is usually read "not p."

Conjunction

The symbol "$\wedge$" stands for *and*. Thus, for any two propositions p and q, the proposition that says that p and q are both true is written $p \wedge q$. It has the following truth table, to reflect the four possible cases: (1) p and q are both true; (2) p is true and q is false; (3) p is false and q is true; (4) p and q are both false.

p	q	$p \wedge q$
T	T	T
T	F	F
F	T	F
F	F	F

Thus $p \wedge q$ is true only in the first of the four possible cases, i.e. when both p and q are true. The proposition $p \wedge q$ is called the *conjunction* of p and q, and is read "p and q."

Disjunction

We write $p \vee q$ to mean that at least one of the two propositions p and q are true, and possibly both.

In ordinary English the word "or" is used in two different senses, in the *strict* or *exclusive sense*, meaning that exactly one of the two choices is true, and in the *inclusive sense*, meaning that *at least* one is true, and possibly both. For example, if I say that tomorrow I will go either East or West, I am using "or" in the exclusive sense, since it is understood that I am not planning to do both. On the other hand, if a university requires that an applicant know either French *or* German, it certainly won't exclude someone who happens to know both, and so "or" is being used there in the inclusive sense.

It is the inclusive sense of "or" that is used in mathematical logic and computer science. The truth table for $\vee$ (which is the logical symbol for "or" that we will use in this text) is thus the following:

p	q	$p \vee q$
T	T	T
T	F	T
F	T	T
F	F	F

Thus $p \vee q$ is false only when both p and q are false. In the other three cases, $p \vee q$ is true. The proposition $p \vee q$ is called the *disjunction* of p and q, and is read "p or q."

The Conditional

The proposition "if p then q," or equivalently, "p implies q," is symbolically rendered $p \supset q$ (sometimes $p \rightarrow q$). In an implication statement $p \supset q$, p is called the *antecedent* and q *the consequent*. We already remarked in Chapter 1 that "if-then," as used in mathematical logic (and also computer science), may well differ in some respects from everyday usage, in that p implies q is to be regarded as false only in the case that p is true and q is false. Thus if p is false, then regardless of whether q is true or false, the statement "p implies q" is to be regarded as true (and of course, if p and q are both true, then p implies q is regarded as true).

In Chapter 1 I gave an example to offer some justification for this seemingly strange use, and here is another: Consider the following proposition about the natural numbers. "If x is odd then $x+1$ is even." Surely you regard that as true, don't you? If you do, then you are committed to believe it even when x is even, and so you are committed to believe the proposition if 4 is odd, then $4+1$ is even. This holds despite the fact that both the antecedent (4 is odd) and the consequent ($4+1$ is even) are false! An so in mathematical logic and computer science, the following is the truth table for $p \supset q$:

p	q	$p \supset q$
T	T	T
T	F	F
F	T	T
F	F	T

Again, the only case in which $p \supset q$ is false is the one in which p is true and q is false.

The Biconditional

We write $p \equiv q$ (sometimes $p \leftrightarrow q$) to mean that p and q are either both true or both false, or, what is the same thing, when it is the case that when either one is true, so is the other, or again, when each implies the other. We read $p \equiv q$ as "p if and only if q," or as "p and q are equivalent" (as far as truth and falsity are concerned). Thus the truth table for the biconditional $\equiv$ is the following:

p	q	$p \equiv q$
T	T	T
T	F	F
F	T	F
F	F	T

Parentheses

One can combine simple propositions into compound ones in many ways, and parentheses are often needed to avoid ambiguities. For example, if we write $p \wedge q \vee r$ it is impossible to tell which of the following is meant:

(1) p is true and $(q \vee r)$ is true.
(2) Either $(p \wedge q)$ is true or r is true.

If we mean (1), we should write $p \wedge (q \vee r)$. If we mean (2), we should write $(p \wedge q) \vee r$.

Formulas

To approach our subject more rigorously, we must define the notion of a *formula*. The letters p, q, r, with or without subscripts, are called *propositional variables*. By a *formula* is meant any expression constructed according to the following rules:

(1) Each propositional variable is a formula.
(2) Given any formulas X and Y already constructed, the expressions $\sim X$, $(X \wedge Y)$, $(X \vee Y)$, $(X \supset Y)$, and $(X \equiv Y)$ are formulas.

It is to be understood that no expression is a formula unless its being so is a consequence of rules (1) and (2).

Exercise. Show that the set of all formulas of Propositional Logic is a denumerable set.

The following definition of matching parentheses may help the reader "parse" any more complex formula that may sometimes appear later in this text, i.e. it may help the reader break the formula down into its meaningful parts, its so-called "sub-formulas," so as to be able to understand the whole formula more clearly.

We can define *matching parentheses* in a formula in two ways (which come down to being the same thing):

(1) If we can see clearly how a formula is built up from its innermost propositional variables and connectives and parentheses, then, at each point that a

pair of parentheses is added around a newly constructed component of the formula, those two parentheses consist in a pair of matching parentheses.

(2) [a] Given a well-formed formula, if a left parenthesis in that formula is followed by a right parenthesis and no parentheses lie between the two, the two parentheses are matching parentheses. [b] Given a well-formed formula, if a left parenthesis in that formula is followed by a right parenthesis and no parentheses lie between the two that have not already been determined to be matching by [a] or [b], then the pair is a pair of matching parentheses.

When two parentheses are matching, we also say *they match each other*. The second definition of matching parentheses above suggests how to pair off the matching parentheses in a complex formula one might be given. One first pairs all parenthesis pairs in which a right parenthesis follows a left parenthesis and there are no parentheses between them. Then one continues pairing the remaining *unmatched* parentheses in the resulting formula in the same way. Incidentally, if, at the end of doing this, some parentheses remain unpaired, that means the original formula was badly formed, i.e. not really a formula of Propositional Logic. At the very least there must be the same number of left and right parentheses in a formula, and one thing formula writers often check is that simple fact, but that's not really enough; ideally one should go through the process of matching each parenthesis with its partner, and check after doing that what's inside any matching pair of parentheses is a good formula.

A way people often write complex formulas to help them to be understandable as well as to help checking that they are well-formed, is to use different kinds of parentheses within the formula, i.e. ordinary parentheses "(and)", square brackets "[and]", curly brackets "{ and }", or even ordinary parentheses of different sizes. These can make matching parentheses more apparent. For instance, consider the following formula:

$$((p \supset (q \supset r)) \supset ((p \supset q) \supset (p \supset r))).$$

Notice now how having different kinds of parentheses at different levels shown with different types of bracketing symbols makes the formula a little more readable (a matching pair of parentheses with no parentheses inside them can be said to be of level 0; if a given matching pair of parentheses has as the highest level pair of matching parentheses inside it a pair of level n, the given pair is of level $n+1$):

$$\{[p \supset (q \supset r)] \supset [(p \supset q) \supset (p \supset r)]\}.$$

Here's the same formula written with parentheses of three different sizes:

$$\Big(\big(p \supset (q \supset r)\big) \supset \big((p \supset q) \supset (p \supset r)\big)\Big).$$

When displaying a formula standing alone, outer parentheses can be deleted without incurring any ambiguity. For example, $(p \wedge \sim\sim q)$ standing alone could be written $p \wedge \sim\sim q$.

Compound Truth Tables

By the *truth value* of a proposition p is meant its truth or falsity, that is, the truth value of p is T if p is true, and is F if p is false. For any two propositions p and q, if we know their truth values, then by the simple truth tables already considered, we can determine the truth value of $\sim p$, $(p \wedge q)$, $(p \vee q)$, $(p \supset q)$, and $(p \equiv q)$. Therefore, given any combination of p and q, that is, given any proposition expressed in terms of p and q using the logical connectives, we can determine the truth value of the combination, given the truth values of p and q. For example, suppose X is the formula $p \equiv (q \vee \sim (p \wedge q))$. Given the truth value of p and q, we can successively determine the truth value of $p \wedge q$, $\sim (p \wedge q)$, $q \vee \sim (p \wedge q)$, and finally of $p \equiv (q \vee \sim (p \wedge q))$. There are four distributions of truth values for p and q: p true, q true; p true, q false; p false, q true; p false, q false.

We can determine the truth value of X in all these cases systematically by constructing the following table, which is an example of a *compoundtruth table*.

p	q	$p \wedge q$	$\sim(p \wedge q)$	$q \vee \sim(p \wedge q)$	$p \equiv (q \vee \sim(p \wedge q))$
T	T	T	F	T	T
T	F	F	T	T	T
F	T	F	T	T	F
F	F	F	T	T	F

We see that X is true in the first two cases and false in the last two.

We can also construct a truth table for a combination of three unknowns, p, q and r, but now there are eight cases to consider, since there are four distributions of T's and F's to p and q, and with each of these four distributions, there are two possibilities for r. As an example, here is a truth table for the formula $(p \vee \sim q) \supset (r \equiv (p \wedge q))$.

p	q	r	$\sim q$	$p \vee \sim q$	$p \wedge q$	$r \equiv (p \wedge q)$	$(p \vee \sim q) \supset (r \equiv (p \wedge q))$
T	T	T	F	T	T	T	T
T	T	F	F	T	T	F	F
T	F	T	T	T	F	F	F
T	F	F	T	T	F	T	T
F	T	T	F	F	F	F	T
F	T	F	F	F	F	T	T
F	F	T	T	T	F	F	F
F	F	F	T	T	F	T	T

For four unknowns, p, q, r, and s, there are $2^4 = 16$ cases to consider, and in general, for any positive integer n, a truth table for a formula with n propositional variables will have 2^n rows corresponding to the 2^n different distributions of truth values to the n variables.

Tautologies

Consider the formula $(p \wedge (q \vee r)) \supset ((p \wedge q) \vee (p \wedge r))$. It has the following truth table:

p	q	r	$q \vee r$	$p \wedge (q \vee r)$	$p \wedge q$	$p \wedge r$	$(p \wedge q) \vee (p \wedge r)$	$(p \wedge (q \vee r)) \supset ((p \wedge q) \vee (p \wedge r))$
T	T	T	T	T	T	T	T	T
T	T	F	T	T	T	F	T	T
T	F	T	T	T	F	T	T	T
T	F	F	F	F	F	F	F	T
F	T	T	T	F	F	F	F	T
F	T	F	T	F	F	F	F	T
F	F	T	T	F	F	F	F	T
F	F	F	F	F	F	F	F	T

We see that the last column contains all T's, which indicates that the formula is true in all eight possible cases. Such formulas are called *tautologies.*

In general, by an *interpretation* of a formula is meant an assignment of truth values (T's and F's) to all the propositional variables in the formula. As we have seen, for a formula with n variables, there are 2^n interpretations, each corresponding to a row in its truth table. A formula is called a *tautology* if it is always true, that, is, true under all its interpretations, or what is the same thing, if the last column of its truth table consists entirely of T's. A formula is called *contradictory* if the last column of its truth table consists entirely of F's, indicating that it is false under all its interpretations. A formula that is neither a tautology nor a contradiction is called *contingent.* Thus a contingent formula is true in some interpretations and false in others, and thus the last column of its truth table has some T's and some F's.

Problem 1. State which of the following are tautologies, which are contradictions, and which are contingent.

(a) $(p \supset q) \supset (q \supset p)$
(b) $(p \supset q) \supset (\sim p \supset \sim q)$

(c) $(p \supset q) \supset (\sim q \supset \sim p)$
(d) $p \supset \sim p$
(e) $p \equiv \sim p$
(f) $(p \equiv q) \equiv (\sim p \equiv \sim q)$
(g) $\sim(p \wedge q) \equiv (\sim p \wedge \sim q)$
(h) $\sim(p \wedge q) \equiv (\sim p \vee \sim q)$
(i) $(\sim p \vee \sim q) \equiv \sim(p \vee q)$
(j) $\sim(p \vee q) \equiv (\sim p \wedge \sim q)$
(k) $(p \equiv (p \wedge q)) \equiv (q \equiv (p \vee q))$

Logical Implication and Equivalence

A formula X is said to logically imply a formula Y if Y is true in all cases in which X is true, or what is the same thing, if $X \supset Y$ is a tautology. Given a set S of formulas and a formula X, we say that X is *logically implied* by S if X is true in all cases in which all elements of S are true. [Later we will learn the interesting fact that for any denumerable set S of formulas, a formula X is logically implied by S if and only if X is logically implied by some finite subset of S! Here is where the denumerable compactness theorem of the last chapter comes in!]

Two formulas X and Y are called *logically equivalent* (sometimes just *equivalent* for short) if they are true in just the same cases, or what is the same thing, if $X \equiv Y$ is a tautology.

Some Abbreviations

It is not difficult to see that for conjunctions and disjunctions, parentheses don't really matter, since $(X_1 \wedge (X_2 \wedge X_3))$ is obviously equivalent to $((X_1 \wedge X_2) \wedge X_3)$ and $(X_1 \vee (X_2 \vee X_3))$ is obviously equivalent to $((X_1 \vee X_2) \vee X_3)$. For each $n > 2$, we abbreviate $((\cdots(X_1 \wedge X_2) \wedge \cdots X_{n-1}) \wedge X_n)$ by $X_1 \wedge X_2 \wedge \cdots \wedge X_n$. For example, we write $X_1 \wedge X_2 \wedge X_3$ to abbreviate $((X_1 \wedge X_2) \wedge X_3)$ and we write $X_1 \wedge X_2 \wedge X_3 \wedge X_4$ for $(((X_1 \wedge X_2) \wedge X_3) \wedge X_4)$, etc. Thus for each $n \geq 2$, we have $X_1 \wedge \cdots \wedge X_n \wedge X_{n+1}$ equal to $(X_1 \wedge \cdots \wedge X_n) \wedge X_{n+1}$. Similar remarks apply to $\vee$ instead of $\wedge$. Thus we have $X_1 \vee \cdots \vee X_n \vee X_{n+1}$ equal to $(X_1 \vee \cdots \vee X_n) \vee X_{n+1}$.

Finding a Formula Corresponding to a Truth Table

Suppose you are given the distributions of T's and F's in the last column of a truth table. Can you find a formula having that column as the last column in its truth table? For example, suppose I consider a case of three variables p, q, and r, and I write down at random T's and F's in the last column like this:

p	q	r	?
T	T	T	T
T	T	F	F
T	F	T	T
T	F	F	F
F	T	T	T
F	T	F	F
F	F	T	F
F	F	F	F

The problem is to find a formula such that the last column of its truth table is the column under the question mark. Do you think that cleverness and ingenuity are required to do this? The answer is that they are not! There is an extremely simple mechanical method that solves all problems of this type. Once you know the method, then regardless of what distribution of T's and F's occur in the last column, you can easily write down the required formula.

Problem 2. What is the method?

Formulas Involving t and f

For certain purposes it is sometimes desirable to add the symbols t and f to the language of Propositional Logic, and extend the notion of *formula* by replacing (1) in the definition by "Each propositional variable is a formula, and so is t and f." Thus, for example, $(p \supset t) \vee (f \wedge q)$ is a formula. The symbols t and f are called *propositional constants* and stand for *truth* and *falsity*, respectively. It is to be understood that in any interpretation t is to be given the value *truth* and f is to be given the value of *falsity*.

Any formula X involving t or f or both, is equivalent either to a formula involving neither, or to t alone, or to f alone. This can be easily seen by virtue of the following equivalences (we abbreviate "is equivalent to" by "equ."

$$
\begin{array}{llll}
X \wedge t & \text{equ } X & t \wedge X & \text{equ } X \\
X \wedge f & \text{equ } f & f \wedge X & \text{equ } f \\
\\
X \vee t & \text{equ } t & t \vee X & \text{equ } t \\
X \vee f & \text{equ } X & f \vee X & \text{equ } X \\
\\
X \supset t & \text{equ } t & t \supset X & \text{equ } X \\
X \supset f & \text{equ } {\sim}X & f \supset X & \text{equ } t
\end{array}
$$

$$X \equiv t \text{ equ } X \qquad t \equiv X \text{ equ } X$$
$$X \equiv f \text{ equ } {\sim}X \qquad f \equiv X \text{ equ } {\sim}X$$

$$\sim t \text{ equ } f \qquad \sim f \text{ equ } t$$

Problem 3. Reduce each of the following formulas either to a formula not involving t or f, or to t alone, or to f alone.

(a) $((t \supset p) \wedge (q \vee f)) \supset ((q \supset f) \vee (r \wedge t))$
(b) $(p \vee t) \supset q$
(c) $\sim(p \vee t) \equiv (f \supset q)$

Liars, Truth Tellers and Propositional Logic

Before continuing with a formal study of Propositional Logic, I would like to interject some problems that illustrate how Propositional Logic is related to recreational logic puzzles about lying and truth-telling of the type considered in many of my earlier puzzle books.

Problem 4. We go back to the land of Problem 1 of Chapter 4, where inhabitants are either of type T or type F, and those of type T make only true statements, while those of type F make only false statements. We consider two inhabitants A and B. A is asked to say something about himself and B.

(a) Suppose A says, "Both of us are of type F." What can be determined about the types of A and B?
(b) Suppose, instead, A says, "At least one of us is of type F." What can then be determined about them?
(c) What if, instead, A says, "B and I are of the same type. We are either both of type T or both of type F." What can then be determined about them?

Interdependence of the Logical Connectives

Some Standard Results

Suppose a man from another planet comes to our planet and wants to study Propositional Logic. He comes to you and says, "I understand the meaning of $\sim$ (not) and of $\wedge$ (and), but I don't understand what the symbol $\vee$ means, and I don't understand what the word "or" means either. Can you explain them to me in terms of $\sim$ and $\wedge$, which I already understand?"

What he wants is a definition of $\vee$ in terms of $\sim$ and $\wedge$, that is, he wants a formula in two propositional variables, say p and q, involving only the connectives $\sim$ and $\wedge$ that is equivalent to the formulas $p \vee q$.

Problem 5. Can you help him out? How do you define $\vee$ in terms of $\sim$ and $\wedge$?

Problem 6. Also, $\wedge$ can be defined in terms of $\sim$ and $\vee$. How?

Problem 7. Define $\supset$ in terms of $\sim$ and $\wedge$.

Problem 8. Define $\supset$ in terms of $\sim$ and $\vee$.

Problem 9. Define $\wedge$ in terms of $\sim$ and $\supset$.

Problem 10. Define $\vee$ in terms of $\supset$ and $\sim$.

Problem 11. Curiously enough, $\vee$ is definable from $\supset$ alone. How? [The solution is quite tricky!]

Problem 12. Define $\equiv$ from $\wedge$ and $\supset$.

Problem 13. Define $\equiv$ from $\sim$, $\wedge$ and $\vee$.

Problem 14. Define $\sim$ from $\supset$ and the propositional constant f (for falsity).

Joint Denial

All the five logical connectives, $\sim$, $\wedge$, $\vee$, $\supset$, $\equiv$, are definable from $\sim$ and $\wedge$ alone (since we have seen that we can first get $\vee$, then $\supset$, and hence $\equiv$). Likewise it follows from the last ten problems that $\sim$ and $\vee$ is a basis for the rest, and so is $\sim$ and $\supset$, as well as $\supset$ and f.

Now there is a single logical connective $p \downarrow q$, called *joint denial* from which all the five connectives $\sim$, $\wedge$, $\vee$, $\supset$, $\equiv$, can be defined. $p \downarrow q$ means "p and q are both false," and is thus equivalent to $\sim p \wedge \sim q$. Its truth table is:

p	q	$p{\downarrow}q$
T	T	F
T	F	F
F	T	F
F	F	T

Problem 15. Prove that from the single connective $\downarrow$ all five of the connectives $\sim$, $\wedge$, $\vee$, $\supset$, $\equiv$, can be obtained. [Hint: First get $\sim$.]

Alternative Denial

There is another single connective from which all the others can be obtained. It is known as *alternative denial*, and also as the *Sheffer Stroke*. $p|q$ is read "at least one of p, q is false, or as "p is false or q is false." It is equivalent both to $\sim p \vee \sim q$ and to $\sim(p \wedge q)$ [which we have already learned are equivalent in Problem 1 (h)]. It has the following truth table:

p	q	$p|q$
T	T	F
T	F	T
F	T	T
F	F	T

Problem 16. Prove that the Sheffer stroke generates all the other connectives.

Remark. It can be proved that no single connective other than $\downarrow$ and $|$ is adequate to yield all the other connectives. Joint denial and alternative denial are the only ones. A proof can be found, e.g. in Mendelson [1987].

Further Results

All the results of interdefinability so far considered here are well known. Less well known are some results of the author, to which we now turn. But first, another liar/truth-teller puzzle.

Problem 17. We return to the land in which every inhabitant is either of type T and makes only true statements, or of type F, and makes only false statements. A prospector goes to this land because of a rumor that gold is buried there. He meets an inhabitant and asks him, "Is there any gold in this land?" The inhabitant replies, "If I am of type T, then there is gold here." What can be determined about the inhabitant's type and about whether or not there is gold there?

The above problem is closely related to the following one:

Problem 18. Show that $\wedge$ is definable from $\supset$ and $\equiv$, and show how this is related to the last problem.

We also have the following further definability problems.

Problem 19. Show that $\vee$ is definable from $\supset$ and $\equiv$.

Problem 20. Show that $\supset$ is definable from $\wedge$ and $\equiv$.

Problem 21. Show that $\wedge$ is definable from $\vee$ and $\equiv$. [This is quite tricky!]

Problem 22. Show that $\vee$ is definable from $\wedge$ and $\equiv$.

Other Connectives

$p \not\equiv q$ (p is not equivalent to q). The truth table of this connective is the following:

p	q	$p \not\equiv q$
T	T	F
T	F	T
F	T	T
F	F	F

Actually, $\not\equiv$ is *exclusive disjunction*, for $p \not\equiv q$ says that one and only one of p, q is true.

Problem 23. Show that $\vee$ is definable form $\wedge$ and $\not\equiv$ (thus inclusive disjunction is definable form conjunction and exclusive disjunction).

Problem 24. Also, $\wedge$ is definable from $\vee$ and $\not\equiv$. Show this.

Problem 25. All the connectives so far considered are definable from $\not\equiv$ and $\supset$. How?

Another connective is $\not\supset$. $p \not\supset q$ is read "p does not imply q" and is thus equivalent to $\sim(p \supset q)$.

Problem 26. Show that $\wedge$ is definable from $\not\supset$.

Problem 27. Show that $\not\supset$ is definable from $\not\equiv$ and $\wedge$.

Problem 28. Show that $\not\supset$ is definable from $\not\equiv$ and $\vee$.

Problem 29. Show that $\not\equiv$ is definable from $\not\supset$ and $\vee$.

Problem 30. Show that all the connectives so far considered are definable from $\not\supset$ and $\supset$.

Problem 31. Show that all the connectives so far considered are definable from $\not\supset$ and $\equiv$.

The Sixteen Logical Connectives

The eight logical connectives between two variables p and q so far considered are the following: $\wedge$, $\vee$, $\supset$, $\equiv$, $\downarrow$, $|$, $\not\equiv$, and $\not\supset$. Let us review their truth tables:

p	q	$p \wedge q$	$p \vee q$	$p \supset q$	$p \equiv q$	$p \downarrow q$	$p \mid q$	$p \not\equiv q$	$p \not\supset q$
T	T	T	T	T	T	F	F	F	F
T	F	F	T	F	F	F	T	T	T
F	T	F	T	T	F	F	T	T	F
F	F	F	F	T	T	T	T	F	F

Two other logical connectives of two variables p and q are $\subset$ (inverse implication) and its negation $\not\subset$. We read $p \subset q$ as "p is implied by q," or as "if q then p." Thus $p \not\subset q$ is read "p is not implied by q."

We now have 10 connectives. There are six more to go, since there are 16 possible truth tables for p and q. These six further connectives are:

p, regardless of q;
not p, regardless of q;
q, regardless of p;
not q, regardless of p;
true, regardless of p and q;
false, regardless of p and q.

p	q	$p \subset q$	$p \not\subset q$	p	q	$\sim p$	$\sim q$	t	f
T	T	T	F	T	T	F	F	T	F
T	F	T	F	T	F	F	T	T	F
F	T	F	T	F	T	T	F	T	F
F	F	T	F	F	F	T	T	T	F

A set of connectives is sometimes called a *basis* (for all the connectives) if all sixteen connectives are definable from them. Now, we have seen that given any truth table, a formula can be found that has it for its truth table – moreover, as seen in the solution of Problem 2, a formula using just the connectives $\sim$, $\vee$, and $\wedge$, so that these three connectives clearly form a basis for all the others. But these three are derived in turn, either from $\sim$ and $\wedge$, or from $\sim$ and $\vee$. Thus both $\sim$, $\wedge$ and $\sim$, $\vee$ are also bases. Other bases are $\sim$ and $\supset$; $\supset$ and f; $\supset$ and $\not\supset$; $\downarrow$ alone; and $|$ alone.

Solutions to the Problems of Chapter 5

1. (a) Contingent.
   (b) Contingent.
   (c) Tautology.
   (d) Contingent. [Note: Many beginners wrongly believe this to be a contradiction. The fact is that when p is false, the formula $p \supset \sim p$ is true. The formula is of course false when p is true.]
   (e) Contradiction.
   (f) Tautology.
   (g) Contingent.
   (h) Tautology.
   (i) Contingent.
   (j) Tautology.
   (k) Tautology.
2. The following solution of the particular case in the text will adequately explain the general method. In this example, the formula we seek is to come out T in the first, third and fifth rows of its truth table. Well, the first row is the case when p, q and r are all true, in other words when $p \wedge q \wedge r$ is true. The third case is when p is true, q is false, and r is true, in other worlds when $p \wedge \sim q \wedge r$ is true. The fifth case is that in which p is false, q is true, and r is true, in other words when $\sim p \wedge q \wedge r$ is true. Thus the formula is to be true when and only when *at least one* of these three cases holds, and so the desired formula is $(p \wedge q \wedge r) \vee (p \wedge \sim q \wedge r) \vee (\sim p \wedge q \wedge r)$.
3. (a) In the formula $((t \supset p) \wedge (q \vee f)) \supset ((q \supset f) \vee (r \wedge t))$ we replace $t \supset p$ by p; $(q \vee f)$ by q; $(q \supset f)$ by $\sim q$; $(r \wedge t)$ by r, thus getting $(p \wedge q) \supset (\sim q \vee r)$.

   (b) In $(p \vee t) \supset q$ we first replace $(p \vee t)$ by t, thus getting $t \supset q$. This further reduces to q.

   (c) In $\sim(p \vee t) \equiv (f \supset q)$. We $(p \vee t)$ by t, then $(f \supset q)$ by t, thus getting $\sim t \equiv t$, thus getting $f \equiv t$ which reduces to f.
4. (a) No one of type T could possibly say that he and someone else were both of type F, for if he did, it would be true that he and the other were of type F, which would make him of both type T and F, which is not possible. Thus the speaker A cannot be of type T, so he must be of type F. Since he is of type F, his statement was false, i.e. the two are not really both of type F. Thus at least one is of type T. Since it is not A, it must be B. Thus the answer is that A is of type F and B is of type T.

   Now let us see how this problem could alternatively be solved by a truth table. The key point to realize about liar/truth-teller puzzles of this type is the following: Suppose an inhabitant of this land asserts a proposition q. Let p be the proposition that the speaker is of type T. Since he asserted q, then if he is of type T, q must be true, but if he is not of type T, then q is false. Thus he is of type T if and only if q is true. Thus if he asserts q, then the reality of the situation is that p is equivalent to q. Thus the fundamental principle is that if he asserts q, then $p \equiv q$ is true!

In this particular problem, A has asserted that he and B are both of type F. Let p be the proposition that A is of type T and q the proposition that B is of type T. A has asserted that p and q are both false, in other words that $\sim p \wedge \sim q$. Since he asserted $\sim p \wedge \sim q$, then the reality of the situation is that $p \equiv (\sim p \wedge \sim q)$ is true. Let us make a truth table for $p \equiv (\sim p \wedge \sim q)$.

p	q	$\sim p$	$\sim q$	$\sim p \wedge \sim q$	$p \equiv (\sim p \wedge \sim q)$
T	T	F	F	F	F
T	F	F	T	F	F
F	T	T	F	F	T
F	F	T	T	T	F

We see that the only case in which the formula comes out T in the last column is the third row of cases, where p is false and q is true. Thus A is of type F and B is of type T.

(b) This time A asserts that at least one of A and B is of type F. If A were of type F, then it would be true that at least one of them was of type F, but those of type F don't make true statements. Hence A must really be of type T. Thus his statement was true, which means that at least one of them is of type F. Since it is not A, it must be B. Thus the answer now is that A is of type T and B is of type F.

Alternatively, solving this by truth-tables, again let p be the proposition that A is of type T and let q be the proposition that B is of type T. Thus A has asserted $\sim p \vee \sim q$, so that the reality of the situation is that $p \equiv (\sim p \vee \sim q)$. If you make a truth table for this formula, you will see that the only case in which there is a T in the last column is that in which p is true and q is false (the second row of cases in the truth table).

(c) No one in this land could possibly claim to be of type F, since if he were of type T, he would not falsely claim to be of type F, and if he were of type F, he would not truthfully claim to be of type F.

Thus also, no one could claim to be of the same type as one of type F, as this would be tantamount to claiming that he himself was of type F. Thus, since A claims to be of the same type as B, then it must be that B is of type T. As for A, it cannot be determined whether he is of type T or type F. It could be that he is of type T, and truthfully claiming to be of the same type as B, or he could be of type F, and thus falsely claiming to be of the same type as B.

It will be particularly instructive to look at this problem from the viewpoint of Propositional Logic. Again let p be the proposition that A is of type T and q

the proposition that B is of type T. This time A is claiming that p is equivalent to q. Since A claims $p \equiv q$, the reality is that $p \equiv (p \equiv q)$. Its truth table is the following:

p	q	$p \equiv q$	$p \equiv (p \equiv q)$
T	T	T	T
T	F	F	F
F	T	F	T
F	F	T	F

Thus the formula comes out true in the first and third rows, the two cases where q is T; in one of those, p is T and in the other p is F. Thus the whole formula is equivalent to q alone.

Another particularly instructive way of looking at it is this: For any three propositions p, q, and r, the proposition $p \equiv (q \equiv r)$ is equivalent to $(p \equiv q) \equiv r$ (as you can easily verify with a truth table). Thus the formula $p \equiv (p \equiv q)$ now under consideration is equivalent to $(p \equiv p) \equiv q$. However, $p \equiv p$ reduces to t, and so the formula reduces to $t \equiv q$, which in turn reduces to q!

5. To say $p \vee q$ is true, i.e. that one of p and q is true, is equivalent to saying that they are not both false. To say that they *are* both false is to say $\sim p \wedge \sim q$; hence to say that this is not the case is to say $\sim(\sim p \wedge \sim q)$. Thus $p \vee q$ is equivalent to $\sim(\sim p \wedge \sim q)$.
6. $p \wedge q$ is true if and only if it is not the case that either one is false. Thus $p \wedge q$ is equivalent to $\sim(\sim p \vee \sim q)$.
7. $p \supset q$ is equivalent to $\sim(p \wedge \sim q)$.
8. $p \supset q$ is equivalent to $\sim p \vee q$.
9. $p \wedge q$ is equivalent to $\sim(p \supset \sim q)$.
10. $p \vee q$ is equivalent to $\sim p \supset q$.
11. $p \vee q$ is equivalent to $(p \supset q) \supset q$
12. $p \equiv q$ is equivalent to $(p \supset q) \wedge (q \supset p)$
13. $p \equiv q$ is equivalent to $(p \wedge q) \vee (\sim p \wedge \sim q)$
14. $\sim p$ is equivalent to $p \supset f$.
15. First we wish to get $\sim p$ from $\downarrow$. Well, $\sim p$ is equivalent to $p \downarrow p$ [$p \downarrow p$ says that p is false and p is false, which is but a repetitious way of saying that p is false.

    Now that we have $\sim$, we can get $\vee$, because $p \vee q$ is equivalent to $\sim(p \downarrow q)$. [To say that $p \vee q$ is true is equivalent to saying that it is not the case that p and q are both false, in other words that not $p \downarrow q$. Reduced to $\downarrow$ alone, $p \vee q$ is equivalent to $(p \downarrow q) \downarrow (p \downarrow q)$.

    Once we have $\sim$ and $\vee$, we can get all the other connectives, $\wedge$, $\supset$ and $\equiv$, by Problems 6, 8 and 13.

16. $\sim p$ is equivalent to $p|p$ [which says p is false or p is false, which more simply says p is false]. Once we have $\sim$, we take $p \wedge q$ to be $\sim(p|q)$ [it is not the case that either p is false or q is false, in other words, p and q are both true]. Once we have $\sim$ and $\wedge$, we can get $\vee$, $\supset$, $\equiv$ by Problems 5, 8, and 13.
17. The speaker asserts that if he is of type T, then there is gold in the land. Suppose he is of type T. Then the statement is true, i.e. there is gold. Thus if he is of type T, then (1) he is of type T; (2) if he is of type T, there is gold. From (1) and (2) it follows that there is gold (assuming he is of type T). This does not prove that there is gold but only that *if* he is of type T, *then* there is old. But he said just that! Thus he made a true statement, hence he is of type T! Thus we now see that he is of type T *and* if he is of type T, then there is gold, from which it follows that there is gold. Thus we see that the speaker is of type T and that there is gold.
18. Let's look at the last problem from the point of view of Propositional Logic. Again let p be the proposition that the speaker is of type T and let q now be the proposition that there is gold in the land. The speaker has asserted that if he is of type T, then there is gold. Thus the speaker has asserted $p \supset q$. Since he asserted that, then the reality of the situation is that $p \equiv (p \supset q)$. A truth table will reveal that $p \equiv (p \supset q)$. is true when and only when p and q are both true, and so $p \wedge q$ is equivalent to $p \equiv (p \supset q)$!

    Without using a truth table, the following alternative argument that $p \equiv (p \supset q)$ is equivalent to $p \wedge q$ is instructive. We are given $p \equiv (p \supset q)$ and we are to infer $p \wedge q$. To do this, we mirror the argument of the preceding problem. Suppose p is true. Since $p \equiv (p \supset q)$ is true, then $p \supset q$ is true. But once we have both p and $p \supset q$, we get q. This proves, not that q is true, but only that *if* p is true, so is q, in other words that $p \supset q$ is true. Thus we have proved $p \supset q$. And since p is equivalent to $p \supset q$ (i.e. $p \equiv (p \supset q)$ is true), we see that p must be true. Thus p is true, and we have seen that $p \supset q$ is true, hence q must be true. Thus $p \wedge q$ is true. [The converse is pretty obvious: If $p \wedge q$ is true, then p and q are both true; hence $p \equiv (p \supset q)$ is true, since both sides of the $\equiv$ are true.] Thus $p \wedge q$ is equivalent to $p \equiv (p \supset q)$.
19. $p \vee q$ is equivalent to $q \equiv (p \supset q)$. You can verify this, as well as the next several solutions, with a truth table.
20. $p \supset q$ is equivalent to $p \equiv (p \wedge q)$.
21. $p \wedge q$ is equivalent to $(p \vee q) \equiv (p \equiv q)$.
22. $p \vee q$ is equivalent to $(p \wedge q) \equiv (p \equiv q)$.
23. $p \vee q$ is equivalent to $(p \wedge q) \not\equiv (p \not\equiv q)$.
24. $p \wedge q$ is equivalent to $(p \vee q) \not\equiv (p \not\equiv q)$.
25. From $\not\equiv$ alone we can get f (falsity) since f is equivalent to $p \not\equiv p$. Then from $\supset$ and f we can get all the other connectives so far considered, as we have seen.
26. $p \wedge q$ is equivalent to $p \not\supset (p \not\supset q)$.
27. $p \not\supset q$ is equivalent to $p \not\equiv (p \wedge q)$.

28. $p \not\supset q$ is equivalent to $q \not\equiv (p \vee q)$.
29. $p \not\equiv q$ is equivalent to $(p \not\supset q) \vee (q \not\supset p)$.
30. From $\not\supset$ alone we can get f, for f is equivalent to $p \not\supset p$. Then from $\supset$ and f we can get all the others, as we have seen.
31. From $\not\supset$ we get f, as in the previous problem. Then from f and $\equiv$ we can get $\sim$, since $\sim p$ is equivalent to $p \equiv f$. Then from $\sim$ and $\not\supset$ we get $\supset$, since $p \supset q$ is equivalent to $\sim(p \not\supset q)$. Now that we have f and $\supset$ (or $\sim$ and $\supset$), we can get all the others.

    Alternatively, we can directly get $\downarrow$ from $\not\supset$ and $\equiv$, since $p \downarrow q$ is equivalent to $q \equiv (p \not\supset q)$ Then from $\downarrow$ we get all the others.

    As another alternative, we can get alternative denial directly from $\equiv$ and $\not\supset$, since $p|q$ is equivalent to $p \equiv (p \not\supset q)$.

6

Propositional Tableaux

We shall later be studying *First-Order Logic*, which is an enormous advance over Propositional Logic, and in fact is an adequate logic for virtually all of mathematics. For First-Order Logic, the method of truth-tables is far from sufficient, and so we now turn to the method known as tableaux (more precisely "analytic tableaux"), which in this chapter will be treated at the elementary propositional level, and which will later on be extended to First-Order Logic.

Let us begin by noting that under any interpretation, the following eight facts hold for any formulas X and Y:

(1) If $\sim X$ is true, then X is false.
If $\sim X$ is false, then X is true.

(2) If $X \wedge Y$ is true, then X and Y are both true.
If $X \wedge Y$ is false, then either X is false or Y is false.

(3) If $X \vee Y$ is true, then either X is true or Y is true.
If $X \vee Y$ is false, then X and Y are both false.

(4) If $X \supset Y$ is true, then either X is false or Y is true.
If $X \supset Y$ is false, then X is true and Y is false.

These are the eight basic facts behind the tableau method for Propositional Logic.

Signed Formulas

For several reasons, it will be useful to incorporate the symbols T and F into our formal language, and define a *signed formula* as an expression $T\,X$ or $F\,X$, where X is a formula as previously defined (and which will now be called an *unsigned formula*). Informally we read "$T\,X$" as "X is true," and "$F\,X$" as "X is false." Under any interpretation of the propositional variables, a signed formula $T\,X$ is called *true* if X is true, and *false* if X is false. A signed formula $F\,X$ is called *true* if X is false, and *false* if X is true.

By the *conjugate* of a signed formula we mean the result of changing "T" to "F" and "F" to "T." Thus $T\,X$ and $F\,X$ are conjugates of each other.

Illustration of the Tableaux Method

Before defining a tableau, we shall illustrate with an example.

Suppose we wish to prove the formula $p \vee (q \wedge r) \supset [(p \vee q) \wedge (p \vee r)]$. We do so with the following tableau, the explanation for which is given immediately following the tableau:

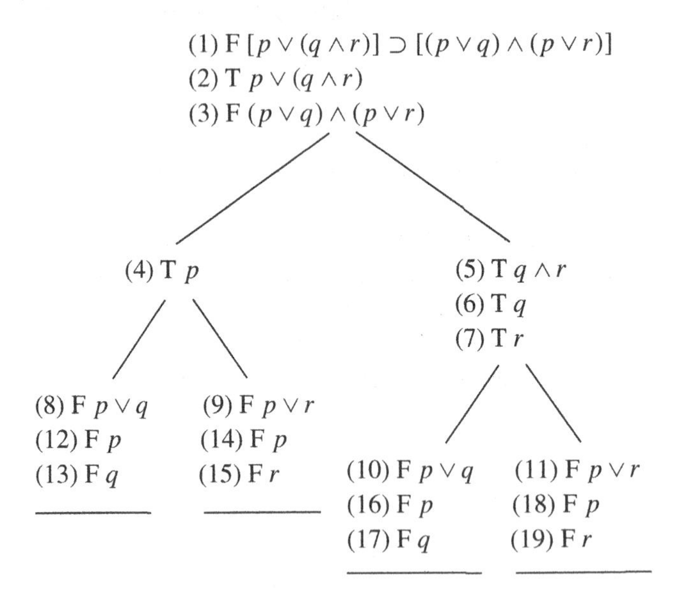

Explanation. The tableau was constructed as follows. As we construct this tableau we are attempting to see if we can derive a contradiction from the assumption that the formula $p \vee (q \wedge r) \supset [(p \vee q) \wedge (p \vee r)]$ is false. And so our first line consists of this formula preceded by the letter "F". Now, a formula of the form $X \supset Y$ can be false only if X is true and Y is false. Thus in the language of tableaux, both T X and F Y are *direct consequences* of the signed formula F $X \supset Y$. Hence we write the lines (2) and (3) as direct consequences of line (1).

Next, we look at line (2), which is of the form T $X \vee Y$, where $X = p$ and $Y = q \wedge r$. We cannot draw any conclusion about the truth of X, nor of the truth of Y. All we can infer is that *either* T X (X is true) or T Y (Y is true). Hence the tableau branches into the two possibilities – lines (4) and (5). Line (5), which is T $q \wedge r$, immediately yields T q and T r as direct consequences, and we thus have lines (6) and (7). Now let's look at line (3). It is of the form F $X \wedge Y$, where $X = p \vee q$ and $Y = \ p \vee r$. No direct conclusion can be drawn; all we can infer is that *either* F X *or* F Y. We also know that either (4) or (5) holds, and so for each of the possibilities (4), (5), we have the two possibilities F X, F Y. Thus there are now four possibilities. And so each of the branches (4), (5) branches again into the possibilities F X, F Y. Thus (4) branches to (8), (9) and (5) branches to (10), (11) [which are, respectively, the same as (8), (9)]. Lines (12), (13) are direct consequences of (8), while lines (14), (15) are direct consequences of (9). Similarly, lines (16) and (17) are direct consequences of (10), and lines (18) and (19) are direct consequences of (11).

The tableau is now *completed*, in the sense that the consequence or consequences of any formula on any line have also been entered into the tableau appropriately. Every line of the tableau that can be used has now been used.

Examining this tableau, let us look at the leftmost branch, which ends with line (13). We can see that lines (4) and (12), two lines on this branch, are direct contradictions of each other [(12) is the *conjugate* of (4)]. Hence they can't both be true, and so we *close* the branch by putting bar under (13), signifying that the branch leads to a contradiction, and so is not a real possibility. Similarly, (14) and (4) contradict each other, and so we can close the branch ending with (15). The next branch (going from left to right) is closed by virtue of (17) and (6). Finally, the rightmost branch is closed by virtue of (19) and (7).

Thus all branches lead to a contradiction, so line (1) is untenable. Thus the formula $p \vee (q \wedge r) \supset [(p \vee q) \wedge (p \vee r)]$ cannot be false under any interpretation, and so is a tautology.

What we do in the tableau method is explore all the possible scenarios in which a formula X could be false. Each branch of a completed tableau shows the situation in one of the possible scenarios. And the reason we could conclude that the formula $p \vee (q \wedge r) \supset [(p \vee q) \wedge (p \vee r)]$ of our example could not ever be false is that we always reached contradictions in following out every single possible scenario.

Remarks

The numbers put to the left of the lines were only for the purpose of identification in the above explanations. We do not need them for the actual construction of tableaux.

We could have closed some of the branches a bit earlier. Lines (13) is really superfluous, since (12) contradicts (4). Hence we could have put a bar right under (12). Similarly we could have closed the branch ending with (15) right after line (14), which contradicts (4). In general, in constructing a tableau, we can close a branch as soon as two contradictory signed formulas appear on it (even if the formulas are complex and contain logical connectives).

We had thus better define a *completed* tableau as one such that on all the *open* branches (i.e. branches that are not closed), all formulas of the branch than can be used *have* been used.

Rules for the Construction of Tableaux

We shall base our propositional tableaux on signed formulas using the logical connectives $\sim$, $\wedge$, $\vee$ and $\supset$. For each of these connectives, there are two rules, one for a formula signed T, and one for a formula signed F. Here are the eight rules. Explanations will follow.

(1) $\dfrac{T \sim X}{F\ X}$ $\qquad$ $\dfrac{F \sim X}{T\ X}$

(2) $\dfrac{T\ X \wedge Y}{T\ X}$ $\quad$ $\dfrac{T\ X \wedge Y}{T\ Y}$ $\qquad$ $F\ X \wedge Y$ branching to $F\ X$ and $F\ Y$

(3) $T\ X \vee Y$ branching to $T\ X$ and $T\ Y$ $\qquad$ $\dfrac{F\ X \vee Y}{F\ X}$ $\quad$ $\dfrac{F\ X \vee Y}{F\ Y}$

(4) $T\ X \supset Y$ branching to $F\ X$ and $T\ Y$ $\qquad$ $\dfrac{F\ X \supset Y}{T\ X}$ $\quad$ $\dfrac{F\ X \supset Y}{F\ Y}$

Some Explanations. Rule (1) means that from T $\sim X$ we can directly infer F X, in the sense that we can subjoin F X to any branch through T $\sim X$. (To "subjoin" here means to add on at the end of the branch.) And from F $\sim X$ we can directly infer T X (again by subjoining it at the end of a branch containing the formula F $\sim X$.) Rule (2) means that T $X \wedge Y$ directly yields both T X and T Y, in the sense that on any branch that contains T $X \wedge Y$ we can subjoin either of the two formulas T X and T Y. On the other hand, Rule (2) also says that if we wish to draw a conclusion from F $X \wedge Y$ on a branch containing that formula, the branch must divide, branching at its end to both F X and F Y. (In other words, in the case of the rule for F $X \wedge Y$, if θ is a branch containing F $X \wedge Y$, when we apply Rule 2 to $F\ X \wedge Y$, we divide θ into two branches at its end, with F X as the last formula of one of the new branches, and F Y as the last formula of the other branch.) Rules (3) and (4) are similarly understood. Note that what we conclude directly from a signed formula according to the tableau rules is precisely everything we can conclude from our assumption about the truth or falsity of the signed formula *based on the meaning of the sign and of the highest level connective in the formula* (of course they can be other connectives in X and Y that went into their formation at an earlier stage). A branch on a tableau is considered *closed* if it contains a formula and its conjugate, and is considered *open* otherwise. A tableau is considered *closed* if all its branches are closed, and is considered open otherwise.

Signed formulas, except for signed variables, are of two types, those of type A, formulas that have one or two direct consequences which we know must be true (T $X \wedge Y$, F $X \vee Y$, T $\sim X$, F $\sim X$) [assuming the formula we are drawing our consequences from is true], and those of type B, formulas that branch to two possible consequences, one of which must be true (F $X \wedge Y$, T $X \vee Y$, T $X \supset Y$).

In constructing a tableau, it is desirable that when a line consisting of a formula of type A occurs, we subjoin both the consequences of the formula to *all* branches that pass through that line. Then that formula of type A need never be used again. In using a line of type B, we divide *all* branches that pass through that line, and then that formula of type B never need be used again. If we construct a tableau in this manner, it is obvious that after a finite number of steps, we reach a point in which every line that can be used has been used, and thus the tableau is complete (because it cannot be further extended without repetitions of conclusions already drawn).

One way of being sure to complete a tableau is to systematically work downward through the tree, i.e. to never use a line until all lines above it (on the same branch) have been used. However, it is more efficient to give priority to lines containing formulas of type A, i.e. to use up all such lines at hand before using lines containing formulas of type B. In this way, one omits repeating the same formula on different branches. Instead, it will have only one occurrence *above* all those branch points. As an example, try the formula $[p \supset (q \supset r)] \supset [(p \supset q) \supset (p \supset r)]$ by making a tableau using each of the above procedures, and you will see that the second one is simpler.

Logical Consequence

The method of tableaux can be used to show that a formula X is a logical consequence of a finite set S of formulas. For example, suppose we wish to show that $p \supset r$ is a logical consequence of the two formulas $p \supset q, q \supset r$. One way, of course, is to construct a closed tableau starting with the formula F $[(p \supset q) \wedge (q \supset r)] \supset (p \supset r)$. But it is more economical (and understandable!) to instead start a tableau with the three lines:

$$\begin{array}{c} \mathrm{T}\, p \supset q \\ \mathrm{T}\, q \supset r \\ \mathrm{F}\, p \supset r \end{array}$$

and show that all branches close.

In general, to show that a formula Y is a logical consequence of formulas $X_1, \ldots, X_n$, we can either construct a closed tableau starting with $(X_1 \wedge \ldots \wedge X_n) \supset Y$, or (and preferably, in terms of the number of times rules must be applied) construct a closed tableau starting with the lines:

$$\begin{array}{c} \mathrm{T}\, X_1 \\ \vdots \\ \mathrm{T}\, X_n \\ \mathrm{F}\, Y \end{array}$$

Tableaux Using Unsigned Formulas

For certain purposes, one of which will be apparent in the next chapter, we will also have use for tableaux using *unsigned formulas*. These differ from those using signed formulas in that "T" is deleted and "F" is replaced by "$\sim$". Thus the rules for constructing tableaus for unsigned formulas are the following [note that the left half of Rule (1) for signed formulas is now superfluous].

(1) $\dfrac{\sim\sim X}{X}$

(2) $\dfrac{X \wedge Y}{X}$ $\quad$ $\dfrac{X \wedge Y}{Y}$ $\quad$ $\sim(X \wedge Y)$ branches to $\sim X$ and $\sim Y$

(3) $X \vee Y$ branches to X and Y $\quad$ $\dfrac{\sim(X \vee Y)}{\sim X}$ $\quad$ $\dfrac{\sim(X \vee Y)}{\sim Y}$

(4) $(X \supset Y)$ branches to $\sim X$ and Y $\quad$ $\dfrac{\sim(X \supset Y)}{X}$ $\quad$ $\dfrac{\sim(X \supset Y)}{\sim Y}$

For tableaux using unsigned formulas, closing a branch means terminating the branch whenever some formula and its *negation* are both on it.

For any unsigned formula Z, it will prove convenient to define $\overline{Z}$ as follows: If Z is of one of the forms $(X \wedge Y)$, $(X \vee Y)$, $(X \supset Y)$, or $(X \equiv Y)$ we take $\overline{Z}$ to be $\sim Z$. If Z is of the form p, where p is a propositional variable, we take $\overline{Z}$ to be $\sim p$. If Z is of one of the forms $\sim X$, $\sim(X \wedge Y)$, $\sim(X \vee Y)$, $\sim(X \supset Y)$, or $\sim(X \equiv Y)$ we take $\overline{Z}$ to be X, $(X \wedge Y)$, $(X \vee Y)$, $(X \supset Y)$ or $(X \equiv Y)$ respectively. $\overline{Z}$ is logically equivalent to $\sim Z$, and is called the *conjugate* of Z.

Proof in the Context of Tableaux for Propositional Logic

We say that we have *proved* that a formula X is a tautology, or simply say we have *proved* the formula X, if we have found a closed tableau starting with F X (in the case of tableaux constructed with signed formulas) or starting with $\sim X$ (in the case of tableaux constructed with unsigned formulas). *Remark.* In the context of unsigned formulas, it is more efficient to start the tableau with $\overline{X}$ (which is logically equivalent to $\sim X$) when X starts with a $\sim$.

As we will see later, it will sometimes be valuable to start a tableau with $T\ X$ (in the case of tableaux constructed with signed formulas) or with X (in the case of unsigned formulas). Such a tableau is called a tableau *for* X. We will also sometimes call a tableau starting with $F\ X$ a tableau for $F\ X$.

Exercise 1. Prove the following formulas by tableaux. [For good practice, prove some using signed formulas and some using unsigned formulas.]

(a) $q \supset (p \supset q)$
(b) $[(p \supset q) \wedge (q \supset r)] \supset (p \supset r)$
(c) $[p \supset (q \supset r)] \supset [(p \supset q) \supset (p \supset r)]$
(d) $[(p \supset r) \wedge (q \supset r)] \supset [(p \vee q) \supset r]$
(e) $[(p \supset q) \wedge (p \supset r)] \supset [p \supset (q \wedge r)]$
(f) $\sim(p \vee q) \supset (\sim p \wedge \sim q)$
(g) $(\sim p \wedge \sim q) \supset \sim(p \vee q)$
(h) $[p \wedge (q \vee r)] \supset [(p \wedge q) \vee (p \wedge r)]$
(i) $[(p \supset q) \wedge (p \supset \sim q)] \supset \sim p$
(j) $[((p \wedge q) \supset r) \wedge (p \supset (q \vee r))] \supset (p \supset r)$

A Unifying Notation

In the tableaux so far considered, we are taking $\sim$, $\wedge$, $\vee$ and $\supset$ as independent logical connectives, which might be called "starters," even though some of them are definable in terms of the others. What we have called "starters" are more usually called "primitive connectives." When we come to axiom systems, some treatments of Propositional Logic, including our own, take the above four as starters. Others take just $\sim$ and $\wedge$, some take $\sim$ and $\supset$, and some take $\sim$, $\wedge$ and $\vee$ as starters. The advantage of using several starters is that proofs *within* the system tend to be shorter and come closer to the way we think informally. Unfortunately, though, when there are several starters, proofs *about* the system, the so-called *metatheory*, which we will soon be studying, tend to be much longer and involve analysis of many different cases.

Another thing: Formulas involving several connectives, which are intuitively quite intelligible, when rewritten as formulas involving fewer connectives, tend to lose their naturalness and aesthetic appeal. Consider, for example the formula known as the syllogism:

$$[(p \supset q) \wedge (q \supset r)] \supset (p \supset r).$$

If we took just $\sim$ and $\wedge$ as starters, the formula would read

$$\sim[[\sim(p \wedge \sim q) \wedge \sim(q \wedge \sim r)] \wedge \sim\sim(p \wedge \sim r)].$$

Which is not a very pretty sight! If further reduced to the single connective $\downarrow$ (the Sheffer stroke), which I don't have the stomach to do, the situation becomes even more horrendous!

Now, I am one who wants to eat his cake and have it too, and so I have elsewhere introduced a unifying notation that combines the advantages of many starters with the advantages of a few. Here it is:

We shall use the Greek letter α (read as "al-pha") to stand for any formula of type A, i.e. any one of the five forms: T $(X \wedge Y)$, F $(X \vee Y)$, F $(X \supset Y)$, T $\sim X$, F $\sim X$. For every such formula α, we define the formulas α_1, α_2, which we call the *components* of α, as follows:

If $\alpha =$ T $(X \wedge Y)$, then $\alpha_1 =$ T X and $\alpha_2 =$ T Y;
If $\alpha =$ F $(X \vee Y)$, then $\alpha_1 =$ F X and $\alpha_2 =$ F Y;
If $\alpha =$ F $(X \supset Y)$, then $\alpha_1 =$ T X and $\alpha_2 =$ F Y;
If $\alpha =$ T $\sim X$, then α_1 and α_2 are both F X;
If $\alpha =$ F $\sim X$, then α_1 and α_2 are both T X.

For perspicuity, we summarize these definitions in the following table:

α	α_1	α_2
T $(X \wedge Y)$	T X	T Y
F $(X \vee Y)$	F X	F Y
F $(X \supset Y)$	T X	F Y
T $\sim X$	F X	F X
F $\sim X$	T X	T X

We note that under any interpretation, any α is true if and only if its components α_1 and α_2 are both true. Accordingly, we refer to α-formulas as ones of *conjunctive type.*

We let β (read as "be′-ta") be any signed formula of type B, i.e. any one of the five forms: F $(X \wedge Y)$, T $(X \vee Y)$, T $(X \supset Y)$, T $\sim X$, F $\sim X$, and we define its *components* β_1 and β_2, as follows:

β	β_1	β_2
F $(X \wedge Y)$	F X	F Y
T $(X \vee Y)$	T X	T Y
T $(X \supset Y)$	F X	T Y
T $\sim X$	F X	F X
F $\sim X$	T X	T X

We note that under any interpretation, β is true if and only if at least one of its components β_1 and β_2 is true. Accordingly, we refer to a β-formula as a formula of *disjunctive type.*

Remark. Of course in constructing a tableau we should treat T $\sim X$ or F $\sim X$ as an α, since it is pointless to split a branch into two identical ones!

For unsigned formulas, we take $X \wedge Y$, $\sim(X \vee Y)$, $\sim(X \supset Y)$, to be of type A, and $X \vee Y$, $\sim(X \wedge Y)$, $X \supset Y$, to be of type B, and we take $\sim\sim X$ to be of both type A and of type B. We take the components as follows:

α	α_1	α_2
$X \wedge Y$	X	Y
$\sim(X \vee Y)$	$\sim X$	$\sim Y$
$\sim(X \supset Y)$	X	$\sim Y$
$\sim\sim X$	X	X

β	β_1	β_2
$\sim(X \wedge Y)$	$\sim X$	$\sim Y$
$X \vee Y$	X	Y
$X \supset Y$	$\sim X$	Y
$\sim\sim X$	X	X

The signed formulas T $\sim X$ and F $\sim X$ (or the unsigned formulas $\sim X$ and X) are the only ones that have been classified as both of type A and of type B. This is because in each one of them, its two components are identical, and therefore to say that both components are true is equivalent to saying that at least one of the components is true.

Using our $\alpha - \beta$ notation, we note the pleasant fact that our eight tableau rules can now be collapsed to two:

Rule A $\quad \dfrac{\alpha}{\alpha_1} \quad \dfrac{\alpha}{\alpha_2}$

Rule B
$$\begin{array}{ccc} & \beta & \\ / & & \backslash \\ \beta_1 & & \beta_2 \end{array}$$

And this is true both for tableaux employing signed and unsigned formulas! Thus we have unified not only the eight rules into two (for each type of tableau, those using signed formulas and those using unsigned formulas), but have also unified our two types of tableaux, something that will be very valuable for our metatheory work.

Note: For unsigned formulas, we defined $\overline{X}$ in such a way that if X is an α (a β), then $\overline{X}$ is respectively a β (an α).

Problem 1. There are other connectives that we could have taken as starters that would fit nicely into our $\alpha - \beta$ scheme, such as $\downarrow$ (joint denial) and $|$ (alternative denial). If we did, what should the tableau rules be for T $X \downarrow Y$, F $X \downarrow Y$,

T $X \mid Y$, F $X \mid Y$? Also, what are their components and which ones are α's and which are β's?

The Bi-Conditional $\equiv$

We have not given tableau rules for $\equiv$, because we could have taken $X \equiv Y$ as an abbreviation of $(X \wedge Y) \vee (\sim X \wedge \sim Y)$, or alternatively, of the equivalent formula $(X \supset Y) \wedge (Y \supset X)$. However, if we work with a formula involving $\equiv$, it is quicker to use the following rules:

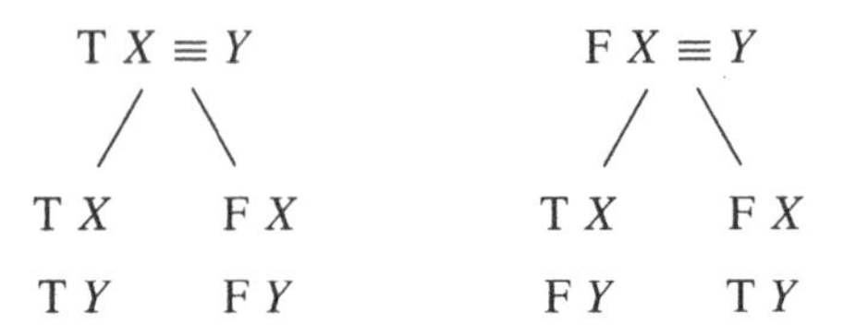

We remark that very few (if any) treatments of Propositional Logic take $\equiv$ as a starter, and also that $\equiv$ does not fit into our $\alpha - \beta$ scheme.

Exercise 2. By tableaux, prove the following:

(a) $[p \equiv (q \equiv r)] \equiv [(p \equiv q) \equiv r]$
(b) $[(p \supset q) \wedge (q \supset p)] \equiv [(p \wedge q) \vee (\sim p \wedge \sim q)]$
(c) $(p \wedge q) \equiv (p \equiv (p \supset q))$
(d) $(p \vee q) \equiv (q \equiv (p \supset q))$
(e) $(p \supset q) \equiv (p \equiv (p \wedge q))$
(f) $(p \wedge q) \equiv [(p \vee q) \equiv (p \equiv q)]$
(g) $(p \vee q) \equiv [(p \wedge q) \equiv (p \equiv q)]$

Degrees

In preparation for the metatheory, which we will soon study, we define the *degree* of an unsigned formula as the number of occurrences of logical connectives in the formula. Thus every propositional variable is of degree 0, and for any formula X and Y of degrees n_1 and n_2, the degree of $\sim X$ is of degree $n_1 + 1$, and each of the formulas $X \wedge Y$, $X \vee Y$, $X \supset Y$ is of degree $n_1 + n_2 + 1$. By the degree of a signed formula T X or F X, we mean the degree of X. Obviously α is of higher degree than both α_1 and α_2, and β is of higher degree than both β_1 and β_2.

Correctness and Completeness

Now we come to the most interesting part, the *metatheory* of tableaux. For simplicity, we will give proofs only for tableaux using signed formulas, unless we say otherwise (but the discussion for tableaux constructed with unsigned formulas would be similar). We wish to prove two things:

(1) *Correctness.* The tableau method is *correct* in the sense that if an unsigned formula X is provable by the tableau method (e.g. if there is a closed tableau starting with FX), then X really is a tautology.

(2) *Completeness.* The tableau method is *complete*, in the sense that every tautology is provable by the tableau method. Better still, if X is a tautoloty, then not only is there a closed tableau starting with FX, but *every* completed tableau starting with FX will be closed.

We remark that the proof of (1) is relatively simple, but the proof of (2) is quite a different story!

Now let us turn to (1): A formula, whether signed or unsigned, is called *satisfiable* if it is true under at least one interpretation. Obviously an unsigned formula X is a tautology if and only if FX is unsatisfiable. A set S of formulas is called *satisfiable* (sometimes *simultaneously satisfiable*) if there is at least one interpretation under which all the elements of S are true. Now, we are to show that for any unsigned formula X, if there is a closed tableau for FX, then X is a tautology, or what is the same, that FX is unsatisfiable.

Let us call a branch of a tableau satisfiable if the set of formulas on the branch is satisfiable, and let us say that a tableau $\mathcal{T}$ is satisfiable if at least one of its branches is satisfiable.

Problem 2.

(a) Show that if a tableau $\mathcal{T}$ is satisfiable, then any extension of $\mathcal{T}$ by an application of Rule A or Rule B is satisfiable.

(b) Then show that if X is any satisfiable (signed) formula, no tableau for X can be closed.

(c) Finally, show that if X is an (unsigned) formula and there is a closed tableau for FX, then X is a tautology, and thus the tableau method is correct.

Now that we have proved that the tableau method is correct, let us turn to the proof that it is complete, i.e. that every tautology is provable by the tableau method ... in fact that if X is a tautology, then every completed tableau for FX must be closed.

Suppose θ is an open branch of a *completed* tableau. Then the set S of formulas on the branch θ satisfies the following three conditions:

H_0: No signed variable and its conjugate are in S. [Actually, no signed formula and its conjugate are in S, but we do not need this stronger fact for what follows.]
H_1: For any α in S, its components α_1 and α_2 are both in S.
H_2: For any β in S, at least one of its components β_1, β_2 is in S.

Sets S, whether finite or infinite, which obey conditions H_0, H_1, H_2, are of fundamental importance and are called *Hintikka sets* (after the logician Jaakko Hintikka, who realized their significance). We now aim to show:

Hintikka's Lemma. Every Hintikka set (whether finite or infinite) is satisfiable.

Problem 3. Prove the following:

(a) Hintikka's Lemma. [Hint: Given a Hintikka set S, you are to describe an interpretation in which all elements of S are true. Let your interpretation be guided by which signed variables are in S. Then, using complete mathematical induction, show that all elements of S are true under that interpretation.]

(b) If a signed formula X is unsatisfiable, then every completed tableau for X must be closed.

(c) If an unsigned formula X is a tautology, then any completed tableau for F X must be closed.

To show that a signed formula T X *is* satisfiable, one constructs a completed tableau, not for F X, but for T X. Any open branch of it will actually yield an interpretation in which T X is true, namely one which assigns *truth* to those variables p for which T p is on the branch, and *falsehood* to those for which F p is on the branch. (For tableaux using unsigned formulas, to show X to be a tautology, one constructs a completed tableau for $\sim X$. To show X satisfiable, one constructs one for X.)

To see whether a finite set $\{X_1, \ldots, X_n\}$ is satisfiable, one constructs a completed tableau for T $X_1 \wedge \cdots \wedge X_n$, or more economically, one starting with

$$\begin{array}{c} \mathrm{T}\ X_1 \\ \mathrm{T}\ X_2 \\ \vdots \\ \mathrm{T}\ X_n \end{array}$$

To see whether a formula Y is a logical consequence of a finite set of formulas $\{X_1, \ldots, X_n\}$, one constructs a completed tableau for F $X_1 \wedge \cdots \wedge X_n \wedge \sim Y$, or more economically, one starting with:

$$\begin{array}{c} \mathrm{T}\ X_1 \\ \mathrm{T}\ X_2 \\ \vdots \\ \mathrm{T}\ X_n \\ \mathrm{F}\ Y \end{array}$$

Compactness

Let us first note that any infinite set of formulas of Propositional Logic must be a denumerable set, since we have already seen (by an Exercise in Chapter 5) that the set of *all* formulas of Propositional Logic is denumerable, which means (by an exercise in Chapter 2) that any infinite subset of the set of all formulas must also be denumerable. Consider now any infinite set S of formulas of Propositional Logic, and, since the set must be denumerable, assume it to be arranged in some sequence $X_1, \ldots, X_n, \ldots$. Suppose there is an interpretation I_1 in which X_1 is true, and there is an interpretation I_2 in which X_1 and X_2 are both true, and in general, for each positive integer n, there is an interpretation I_n in which the first n formulas $X_1, \ldots, X_n$, are all true. Does there necessarily exist an interpretation I in which *all* the infinitely many formulas of S are true? This question is really equivalent to the following question: If all finite subsets of our set S are satisfiable, is the whole of S (simultaneously) satisfiable?

Problem 4. Show that the two questions are equivalent, i.e. show that every finite subset of S is satisfiable if and only if for every n, the set $\{X_1, \ldots, X_n\}$ is satisfiable.

We shall now prove that the answer to the original question is affirmative! That is, we will show that if S is an infinite set of formulas of Propositional Logic all of whose finite subsets are satisfiable, then S itself is also satisfiable. This is known as the *Compactness Theorem for Propositional Logic*, and will be of great use when we come to First-Order Logic.

We give two different proofs of this important result. The first uses tableaux and König's Lemma (or the Fan Theorem). The other uses neither tableaux nor König's Lemma (nor the Fan Theorem), but instead uses the fact, proved in the last chapter, that for any denumerably infinite set S, and any compact property P of subsets of S, any subset A of S having property P can be extended to (i.e. is a subset of) a *maximal* set having property P. Both proofs are of interest, and each reveals certain interesting facts not revealed by the other.

For the first proof, consider a set S of formulas such that all finite subsets are satisfiable. Arrange the elements of S in some infinite sequence $X_1, \ldots, X_n, \ldots$. Now construct a tableau as follows: Run a completed tableau for X_1. It cannot be closed, since X_1 is satisfiable. Now tack on X_2 to the end of every open branch and continue the branch to completion. This new, extended tableau cannot be closed either, since the set $\{X_1, X_2\}$ is satisfiable. Then tack on X_3 to the end of every open branch and run the branch to completion. Again, it cannot be closed, since $\{X_1, X_2, X_3\}$ is satisfiable. Continue this process indefinitely, successively tacking on $X_4, X_5, \ldots, X_n, \ldots$. You thus generate an infinite tableau. At no stage can the tableau close, since for every n, the set $\{X_1, \ldots, X_n\}$ is satisfiable. The tree is obviously finitely generated, since each junction point has at most two successors (one, if it is an α, two, if it is a β). Then by König's Lemma, the tableau has an infinite branch θ. The branch θ is not closed, since every closed branch is finite. And every element of S has been placed on the branch θ at some point. So the set S' of all formulas on the branch θ is a Hintikka set, and is thus satisfiable according to Hintikka's Lemma. Since S is a subset of S', then S is obviously satisfiable. This concludes the proof.

Now for the second proof, which uses neither tableaux nor König's Lemma.

Truth Sets

Let us first consider signed formulas. For any interpretation I, by the *truth set* of I, we shall mean the set of all signed formulas that are true under I. We shall call a set S of signed formulas a *truth set* if it is the truth set of some interpretation I. If S is a truth set, then for any signed formula X and any α and β, the following three conditions hold.

T_0: Either X or its conjugate $\overline{X}$ is in S, but not both.
T_1: α is in S if and only if α_1 and α_2 are both in S.
T_2: β is in S if and only if either β_1 is in S or β_2 is in S.

We will soon see that any set S satisfying T_0, T_1, T_2, is a truth set.

Full Sets

We shall call a set S of signed formulas *full* if, for every signed formula X, either X or its conjugate $\overline{X}$ is in S.

Problem 5. Prove each of the following.

(a) If S is both satistiable and full, then S is a truth set.
(b) Every full Hintikka set is a truth set.
(c) Any set S satisfying conditions T_0, T_1, and T_2 is a truth set.

Exercise 3. There is redundancy in conditions T_0, T_1, and T_2. Condition T_2 follows from T_0 and T_1. Why? Also, Condition T_1 follows from T_0 and T_2. Why? Prove also that if we weaken T_0 to just the case that X is a signed variable, then S is still a truth set.

In preparation for our second proof of the Compactness Theorem for Propositional Logic, the following lemma is key:

Lemma. For any set S of signed formulas and any signed formula X, if some finite subset of $S \cup \{X\}$ is unsatisfiable, and if some finite subset of $S \cup \{\overline{X}\}$ is unsatisfiable, then some finite subset of S is unsatisfiable.

Note: The above lemma can be equivalently stated as follows: If all finite subsets of S are satisfiable, then either all finite subsets of $S \cup \{X\}$ are satisfiable, or all finite subsets of $S \cup \{\overline{X}\}$ are satisfiable.

Problem 6. Prove the above lemma.

Now for the second proof of the compactness theorem for Propositional Logic:

We recall from Chapter 4 that for any property P of subsets of a set A, we defined $P^*(S)$ (where S is a subset of A) to mean that all finite subsets of S have property P. And we recall the Denumerable Compactness Theorem, which is that for any denumerable set A and any compact property P of subsets of A, any subset S of A having property P is a subset of a maximal subset S^* of A having property P.

We now take A to be the set of all signed formulas and P to be the property of satisfiability. Then the property P^* of a set S of signed formulas is that all finite subsets of S are satisfiable. Well, suppose S is denumerable and all finite subsets of S are satisfiable. Then by the Denumerable Compactness Theorem (and the fact that we have shown P^* to be compact for any property P), S is a subset of some *maximal* set S^* of formulas having the property that all finite subsets of S^* are satisfiable. We must now show that S^* is a truth set.

Problem 7. Prove that S^* is a truth set.

Now that we know that S^* is a truth set, then of course S^* is satisfiable, hence its subset S is satisfiable, which concludes our second proof of the Compactness Theorem for Propositional Logic.

We recall that we are calling a formula X a *logical consequence* of a set S of formulas when X is true under any interpretation in which all members of S are true.

The Compactness Theorem for Propositional Logic has the following corollary:

Corollary. If X is a logical consequence of S then X is a logical consequence of some finite subset of S.

Problem 8. Prove the above corollary.

The Compactness Theorem for Propositional Logic also holds for sets of unsigned formulas. The proofs are the same, everywhere replacing T X by X and FX by $\sim X$ or $\overline{X}$. Hence the corollary also holds, and so if X is a logical consequence of S, then X is a logical consequence of some finite subset $\{X_1, \ldots, X_n\}$ of S, which means that the formula $(X_1 \wedge \cdots \wedge X_n) \supset X$ is a tautology.

Dual Tableaux

In a paper "Duals of Smullyan Trees" [1972], the authors Hugues Leblanc and D. Paul Snyder introduced what I will call *dual tableaux* (which they called *dual Smullyan trees*). These will have a useful application in the next chapter.

In uniform notation, the dual tableau rules are the following:

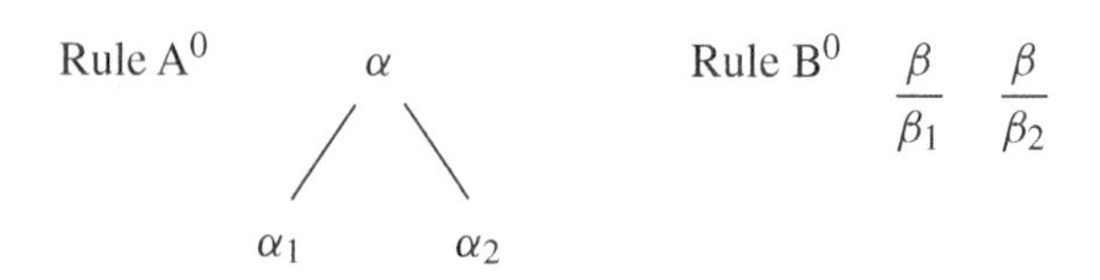

Note. There are dual tableaux both for signed and unsigned formulas. [The authors of "Duals of Smullyan Trees" worked only with unsigned formulas.] For dual tableaux employing signed formulas, we shall regard formulas T $\sim X$, F $\sim X$ as β's (of Type B). For unsigned formula we take $\sim\sim X$ to be of type B.

It should be helpful to realize that under any interpretation I, an α is *false* if and only if either α_1 is false or α_2 is false. And a formula β is false if and only if β_1 and β_2 are both false. These facts are what dual tableaux explore.

For now, we shall consider dual tableaux using signed formulas.

To avoid possible confusion, what we have previously called a *tableau* will presently be called an *analytic tableau* (as in Smullyan, 1978), as distinguished from a dual tableau. Now, an analytic tableau proof of an unsigned formula X is a closed tableau for F X. We now define a *dual tableau proof of* X to be a closed dual tableau for T X. We wish to prove that the dual tableau method is both correct and complete in the sense that there exists a dual tableau proof of X if and only if X is a tautology. We could prove this from scratch, as did Leblanc and Snyder in "Duals of Smullyan Trees" [1973], but it is much quicker to establish this result as a corollary of the correctness and completeness of the analytic tableau method, which we have already proved.

For any tableau $\mathcal{T}$, whether analytic or dual, by its *conjugate tableau* $\overline{\mathcal{T}}$ we shall mean the result of replacing each formula X that occurs anywhere in $\mathcal{T}$ by its conjugate $\overline{X}$. Thus to get from $\mathcal{T}$ to $\overline{\mathcal{T}}$ we simply replace every T by F and every F by T.

As for *correctness*, suppose $\mathcal{T}$ is a closed dual tableau for T X. Then its conjugate $\overline{\mathcal{T}}$ is an analytic tableau for F X and is also closed (why?). Then by the correctness of the analytic tableau method, the signed formula F X is unsatisfiable, hence X is a tautology. Thus the dual tableau method is correct.

Now, suppose X is a tautology. Then there is a closed analytic tableau $\mathcal{T}$ for F X, by the completeness of the analytic tableau method, which we have already proved. Hence the conjugate $\overline{\mathcal{T}}$ of $\mathcal{T}$ is a closed dual tableau for T X, and thus a dual tableau proof of X. And so every tautology has a dual tableau proof, and thus the dual tableau method is complete.

As for dual tableaux using unsigned formulas, if X is a tautology, then there is a closed dual tableau $\mathcal{T}$ starting with T X. If we delete all the T's and replace each F by "$\sim$", we then have a dual tableau proof of X using unsigned formulas. Thus every tautology has a dual tableau proof using unsigned formulas, and this is a result that we will need for the next chapter.

Further insight into dual tableaux should be provided by the following exercise.

Exercise 4. This exercise has the following four parts:

(a) Along the lines of the Leblanc-Snyder article [1973], define a *dual Hintikka set* (called a *dual model set* the paper referred to) as a set S of signed formulas such that for every α, β and X:

$H_0{}^0$: T X and F X are not both in S.
$H_1{}^0$: If α is in S, then either α_1 or α_2 is in S.
$H_2{}^0$: If β is in S, then both β_1 and β_2 are in S.

Prove that if S is a dual Hintikka set, then there is an interpretation in which all elements of S are false.

(b) Now, using (a), give a *direct proof* – one that does not use the completeness result for analytic tableaux – that for every tautology X, there is a closed dual tableau for $T\ X$.

Exercise 5. Call a tableau, whether analytic or dual, *atomically closed*, if for every branch there is a propositional *variable* p such that $T\ p$ and $F\ p$ are on the branch. Prove that if X is a tautology, then there exists an *atomically closed* analytic tableau for FX and an atomically closed dual tableau for $T\ X$.

Solutions to the Problems of Chapter 6

1. The tableau rules for $\downarrow$ and $|$ should be:

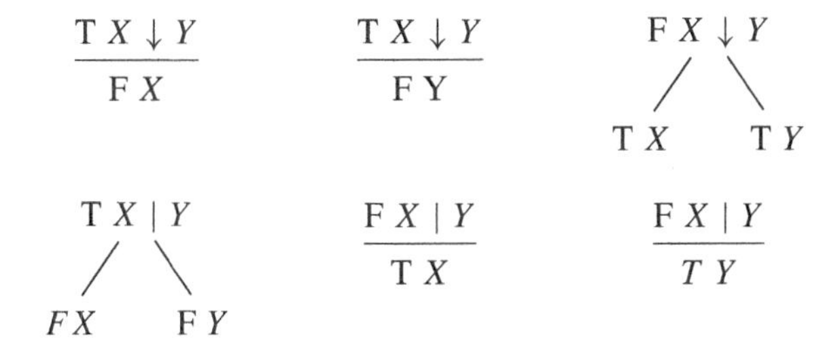

T $X \downarrow Y$ and F $X \mid Y$ are α's and F $X \downarrow Y$ and T $X \mid Y$ are β's.

2. Solutions to the three parts:

(a) Suppose the tableau $\mathcal{T}$ is satisfiable. Let θ be the branch of $\mathcal{T}$ that we extend by Rule A or Rule B. Either θ is satisfiable or it isn't. If it isn't, then some other branch γ of $\mathcal{T}$ is satisfiable. Then extending θ has no effect on the branch γ, so that γ remains satisfiable, and thus the extended tableau – call it $\mathcal{T}_1$ – is satisfiable. Now consider the more interesting case in which θ is satisfiable. Let I be an interpretation under which all the formulas on θ are true. Suppose we extend θ by Rule A. There is some α on the branch θ and we have added either α_1 or α_2 to the branch. Since α is true under I, so are both α_1 and α_2. Thus, whichever one we added, the resulting branch will be satisfiable. Now suppose we extend θ by Rule B. Then some β is on the branch θ, and θ has been split into two branches θ_1 and θ_2, where θ_1 is the result of adding β_1 and θ_2 is the result of adding β_2. Since β is true under the interpretation I, either β_1 or β_2 is true under the interpretation I as well, so that all elements of either θ_1 or θ_2 are true. Thus either θ_1 or θ_2 is satisfiable, and so the extended tableau $\mathcal{T}_1$ is satisfiable.

(b) Let X be a satisfiable signed formula and start a tableau $\mathcal{T}$ with the single formula X. Clearly the tableau $\mathcal{T}$ is satisfiable. Then by (a) any immediate extension $\mathcal{T}_1$ of the tableau $\mathcal{T}$ is satisfiable, hence any immediate extension $\mathcal{T}_2$ of $\mathcal{T}_1$ is satisfiable, and so forth. Since every extension of $\mathcal{T}$ is satisfiable, no extension of $\mathcal{T}$ can be closed (since a closed tableau is obviously not satisfiable).

(c) In particular, for any unsigned formula X, if F X is satisfiable, then no tableau for F X can be closed. Stated otherwise, if there is a closed tableau for F X, then F X is not satisfiable, which means that X is a tautology.

3. Solution to the three parts:

(a) Let S be a set of formulas that satisfies the conditions for a Hintikka set. For every propositional variable p that occurs in any element of S, if $T\ p$ is an element of S, we obviously give p the value truth, and if $F\ p$ is an element of S, we give p the value of falsehood. [By H_0, $T\ p$ and $F\ p$ are not both in S, which guarantees that we can do this with no difficulty.] If neither T p nor F p is an element of S, it makes no difference what value we give to p. We then have an interpretation I in which all elements of degree 0 (all signed propositional variables) that are in S are true. We show by complete mathematical induction on the degrees of the formulas in S that all elements of S are true under I.

Suppose n is a number such that all formulas in S of degree n or less are true under I. (The smallest number the degree n can be is zero, when the formula contains no logical connectives, i.e. is a signed propositional variable. The statement is clearly true in this case.) We must show that under the induction assumption just stated, the statement holds for all elements of S of degree $n+1$. And so let X be an element of S of degree $n+1$. Because the smallest value of $n+1$ is 1, X has at least one logical connective, so is either an α or a β. Suppose it is an α. Then α_1 and α_2 are both in S (by condition H_1), and are both of lower degree than $n+1$, hence are of degree

n or less, and therefore are true by the induction hypothesis. Since α_1 and α_2 are both true, so is α.

On the other hand, suppose X is some β. Then by H_2, either β_1 or β_2 is in S. Whichever one is in S, it is of lower degree than $n+1$, hence is of degree n or less, and therefore true by the indication hypothesis. Thus either β_1 or β_2 is true, which makes β true. This concludes the induction.

(b) Let X be an unsatisfiable signed formula and consider any completed tableau starting with X. Since every open branch of a completed tableau is a Hintikka set, and every Hintikka set is satisfiable, then any open branch of a completed tableau is satisfiable (i.e. the set of formulas on the branch is satisfiable). But since our unsatisfiable X is on every branch of any completed tableau of which it is the topmost formula, no completed tableau for X can have an open branch. Thus every completed tableau for FX must be closed.

(c) If now the unsigned formula X is a tautology, then FX is not satisfiable, hence by (b) every completed tableau for FX must be closed.

4. One direction is obvious. If every finite subset of S is satisfiable, then of course for every n, the set $\{X_1, \ldots, X_n\}$, being finite, is satisfiable.

Conversely, suppose that for every n, the set $\{X_1, \ldots, X_n\}$ is satisfiable. Now consider any finite subset A of S. Let n be the greatest number such that $X_n \in A$. Then A is a subset of $\{X_1, \ldots, X_n\}$, and since this set is satisfiable, so is its subset A. [Of course if A is the empty set, then A is vacuously satisfiable – indeed all elements of A are true under all interpretations, since no elements are in A.]

5. Solution to the three parts:

(a) Suppose S is satisfiable and full. Since S is satisfiable, there is an interpretation I under which all elements of S are true. We wish to show that every formula X that is true under I is in S, and thus that S is the truth set of I.

Suppose X is true under I. Then $\overline{X}$ is not true under I, so that $\overline{X}$ cannot be in S (since only elements that are true under I are in S). Since $\overline{X}$ is not in S and S is full, then $X \in S$. This completes the proof.

(b) This follows from (a), since by Hintikka's Lemma, every Hintikka set is satisfiable.

(c) Any set satisfying T_0, T_1, and T_2 obviously satisfies H_0, H_1, and H_2, and is thus a Hintikka set. It must also be full by T_0, hence is a truth set by (b).

6. We are given that some finite subset of $S \cup \{X\}$ is unsatisfiable. Such a set is either $S_1 \cup \{X\}$ for some finite subset S_1 of S, or is itself a finite subset S_1 of S, in which case $S_1 \cup \{X\}$ is also unsatisfiable. Thus, in either case, there is a finite subset S_1 of S such that $S_1 \cup \{X\}$ is unsatisfiable. Similarly, since we are also given that some finite subset of $S \cup \{\overline{X}\}$ is unsatisfiable, then there is a finite subset S_2 of S such that of $S_2 \cup \{X\}$ is unsatisfiable. Let S_3 be the union of S_1 and S_2 (which we recall is the set of all elements of S_1 together with all elements of S_2). Since S_1 is a subset of S_3, then $S_1 \cup \{X\}$ is a subset of $S_3 \cup \{X\}$, and

since $S_1 \cup \{X\}$ is unsatisfiable, so is $S_3 \cup \{X\}$. Similarly, since S_2 is a subset of S_3, then $S_2 \cup \{\overline{X}\}$ is a subset of $S_3 \cup \{\overline{X}\}$, and since $S_2 \cup \{\overline{X}\}$ is unsatisfiable, so is $S_3 \cup \{\overline{X}\}$. Thus $S_3 \cup \{X\}$ and $S_3 \cup \{\overline{X}\}$ are both unsatisfiable, so that S_3 must be unsatisfiable (because if it were satisfiable, all elements of S_3 would be true under some interpretation I, and either X or $\overline{X}$ would be true under I, and thus either $S_3 \cup \{X\}$ or $S_3 \cup \{\overline{X}\}$ would be satisfiable, which is not the case). Thus the finite subset S_3 of S is unsatisfiable.

7. First we prove that the set S^* is full. To begin with, let us note a certain general fact. Suppose M is a maximal subset of a set A such that M has a certain property P. Then for any element X of A, if $M \cup \{X\}$ has property P, then X must be in M, since if it were not, then M would be a *proper* subset of the set $M \cup \{X\}$ having property P, contrary to the fact that M is a *maximal* subset of A that has property P.

   Now, S^* is a maximal set having the property that all its finite subsets are satisfiable. Then for any formula X, if all finite subsets of $S^* \cup \{X\}$ are satisfiable, then X must be a member of S^*. By the lemma proved in Problem 6, either all finite subsets of $S^* \cup \{X\}$ are satisfiable or all finite subsets of $S^* \cup \{\overline{X}\}$ are satisfiable. In the first case, as we have seen, X must be in S^*; in the second case, $\overline{X}$ must be a member of S^*. Thus either $X \in S^*$ or $\overline{X} \in S^*$, which means that S^* is full.

   Now we show that S^* is a Hintikka set.

   Regarding H_0. If X and $\overline{X}$ were both in S^*, then the finite subset $\{X, \overline{X}\}$ of S^* would be unsatisfiable, contrary to the fact that all finite subsets of S^* are satisfiable. Hence X and $\overline{X}$ cannot both be in S^*, which proves H_0.

   Regarding H_1. Suppose α in S^*. Then $\overline{\alpha_1}$ cannot be in S^*, since the finite subset $\{\alpha, \overline{\alpha_1}\}$ of S^* is not satisfiable. Since $\overline{\alpha_1} \notin S^*$ and S^* is full, then $\alpha_1 \in S^*$. Similarly, $\alpha_2 \in S^*$.

   Regarding H_2. Suppose β in S^*. Then $\overline{\beta_1}$ and $\overline{\beta_2}$ cannot both be in S^*, since the finite subset $\{\beta, \overline{\beta_1}, \overline{\beta_2}\}$ of S^* is not satisfiable. Hence either $\overline{\beta_1} \notin S^*$ or $\overline{\beta_2} \notin S^*$. Since S^* is full, in the first case, $\beta_1 \in S^*$ and in the second case $\beta_2 \in S^*$. Thus $\beta_1 \in S^*$ or $\beta_2 \in S$, which proves H_2.

   Thus S^* is a Hintikka set, and since S^* is also full, S^* is a truth set by (b) of Problem 5. This concludes the proof.

8. To say that X is a logical consequence of S is equivalent to saying that $S \cup \{\overline{X}\}$ is unsatisfiable.

   Now, suppose X is a logical consequence of S. Then $S \cup \{\overline{X}\}$ is unsatisfiable. Hence by the Compactness Theorem for Propositional Logic, some finite subset of $S \cup \{\overline{X}\}$ is unsatisfiable. Thus there is a finite subset S_1 of S such that $S_1 \cup \{\overline{X}\}$ is unsatisfiable, and so X is a logical consequence of S_1.

7
Axiomatic Propositional Logic

The earlier treatment of Propositional Logic was through axiom systems. Truth tables came much later, and tableaux later still. The axiomatic treatment has certain advantages that we will see in later chapters, besides the fact that axiom systems are quite interesting in their own right!

Axiom systems, in the ancient Greek sense of the term, consisted of a set of propositions considered *self-evident*, called *axioms*, and a set of apparently unquestionably valid logical rules that enabled one to derive from these self-evident propositions other propositions, many of which were far from self-evident.

In the modern sense of the term, an axiom system consists of a set of formulas called the *axioms* of the system, together with certain relations called *inference rules*, each being of the form "From the formulas $X_1, \ldots, X_n$, one can infer formula X." By a *proof* in the axiom system is meant a finite sequence of formulas, usually displayed vertically rather than horizontally, called the *lines* of the proof, such that each line is either an axiom or is inferable from earlier lines by one of the inference rules. A formula X is called *provable* in the system if it is the last line of a proof, and such a proof is called a *proof of* X.

This notion of *proof*, unlike the informal notion of proof, is purely objective. It is easy to program a computer to decide whether a sequence of lines is really a proof in a given axiom system or not.

An inference rule "From $X_1, \ldots, X_n$, to infer X" is usually displayed diagrammatically like this:

$$\frac{X_1, \ldots, X_n}{X}$$

A standard rule for Propositional Logic is the rule known as *Modus Ponens:*

$$\frac{X, \ X \supset Y}{Y}$$

This rule can be read as "From the formulas X and $X \supset Y$, the formula Y can be inferred."

We note that this rule is *correct*, in the sense that if X and $X \supset Y$ are both tautologies, so is Y.

The older axiom systems for Propositional Logic took a finite number of tautologies as axioms and used two inference rules, Modus Ponens and another rule called the *Rule of Substitution*, which I will now explain. By an *instance* (sometimes called a *substitution instance*) of a formula X is meant any formula obtained by substituting formulas for some or all of the propositional variables in X. For example, in the formula $p \supset p$, if we substitute $(p \wedge \sim q)$ for p, we obtain the instance $(p \wedge \sim q) \supset (p \wedge \sim q)$. We note that since $p \supset p$ is true for every possible truth value of p, it is true when p is the formula $(p \wedge \sim q)$. Thus, since $p \supset p$ is a tautology, so is its instance $(p \wedge \sim q) \supset (p \wedge \sim q)$. In general, any instance of a tautology is a tautology.

The *Rule of Substitution* is "From X to infer any instance of X." As I said, the earlier axiom systems for Propositional Logic took a finite set of tautologies as axioms and had Modus Ponens and the Rule of Substitution as the two (and only) inference rules. Then, to repeat a point, a proof in these systems consisted of a finite sequence of formulas such that each term of the sequence was either an axiom, or a substitution instance of an earlier term in the sequence, or was inferable from two earlier terms by Modus Ponens.

I personally find it remarkable that *all* tautologies, of which there are infinitely many, can be derived from a *finite* set of tautologies using just the rules of Substitution and Modus Ponens!

Most of these systems are really works of art! In setting up these systems, one takes certain of the logical connectives as starters, more technically called *primitive connectives*, or *undefined connectives*, and defines other connectives in terms of them in the manner explained in Chapter 4. The system used by Russell and Whitehead in *Principia Mathematica* [1910] used $\sim$ and $\vee$ as primitives (and took $X \supset Y$ as merely an abbreviation for $\sim X \vee Y$). The system had five axioms:

(a) $(p \vee p) \supset p$
(b) $p \supset (p \vee q)$
(c) $(p \vee q) \supset (q \vee p)$
(d) $(p \supset q) \supset ((r \vee p) \supset (r \vee q))$
(e) $(p \vee (q \vee r)) \supset ((p \vee q) \vee r)$

It later turned out that axiom (e) was redundant – it was derivable from the other four axioms.

There are many axioms systems that take $\sim$ and $\supset$ as primitives. The earliest such system goes back to Gottlob Frege [1879], and consisted of six axioms. Later, Jan Łukasiewicz [selected writings: 1970] replaced those six axioms by a simpler system that used the following three axioms:

(1) $p \supset (q \supset p)$
(2) $[p \supset (q \supset r)] \supset [(p \supset q) \supset (p \supset r)]$
(3) $(\sim p \supset \sim q) \supset (q \supset p)$

A more modern type of axiom system, which I believe is due to John von Neumann, took as axioms, instead of a finite number of formulas, *all instances* of a finite number of formulas, and used Modus Ponens as the only inference rule. A formula, all of whose instances are taken as axioms, is called an *axiom scheme*. Thus in more modern systems, one takes a finite number of axiom *schemes*, and uses only Modus Ponens as the inference rule. Thus the axioms of a modernized version of the above axiom system consists of formulas of any one of the following three forms:

i. $X \supset (Y \supset X)$
ii. $[X \supset (Y \supset Z)] \supset [(X \supset Y) \supset (X \supset Z)]$
iii. $(\sim X \supset \sim Y) \supset (Y \supset X)$

Each of the i, ii, and iii above is an axiom scheme. Thus the above axiom system has infinitely many axioms, but only three axiom schemes.

A variant of the above system replaces scheme iii by the following scheme:

iii′. $(\sim Y \supset \sim X) \supset [(\sim Y \supset X) \supset Y]$

The completeness of this variant of an axiom system for Propositional Logic is a bit quicker to establish. [An axiom system for Propositional Logic is called *complete* if all tautologies are provable in the system.]

Alonzo Church [1956] gives an axiom system that takes only $\supset$ and f as primitive. In it, $\sim X$ is defined as $X \supset f$. This system used schemes i and ii above, and in place of iii used

iii″. $((X \supset f) \supset f) \supset X$.

J. Barkley Rosser [1953] took $\sim$ and $\wedge$ as primitives. He took $X \supset Y$ as an abbreviation of $\sim(X \wedge \sim Y)$, and used the following three axiom schemes:

(a) $X \supset (X \wedge X)$
(b) $(X \wedge Y) \supset X$
(c) $(X \supset Y) \supset [\sim(Y \wedge Z) \supset \sim(Z \wedge X)]$

Axiom systems using only one axiom scheme have been given. One such axiom system uses $\sim$ and $\supset$ a primitives. Another uses only the one connective $|$ (alternative denial) as primitive. The interested reader can find these schemes in Elliot Mendelson [1987] or Church [1956].

In a book of Stephen Kleene [1952] a system is described that takes all four connectives $\sim$, $\wedge$, $\vee$, $\supset$ as primitives and has the following ten schemes:

K_1: $X \supset (Y \supset X)$
K_2: $(X \supset Y) \supset [(X \supset (Y \supset Z)) \supset ((X \supset Y) \supset (X \supset Z))]$
K_3: $X \supset (Y \supset (X \wedge Y))$
K_4: $(X \wedge Y) \supset X$
K_5: $(X \wedge Y) \supset Y$

K_6: $X \supset (X \vee Y)$
K_7: $Y \supset (X \vee Y)$
K_8: $(X \supset Z) \supset [(Y \supset Z) \supset ((X \vee Y) \supset Z)]$
K_9: $(X \supset Y) \supset ((X \supset \sim Y) \supset \sim X)$
K_{10}: $\sim\sim X \supset X$

I shall call this system *System* $\mathcal{K}$. As I remarked in the last chapter, systems using several primitives have the advantage that proofs within the system tend to be more natural.

The axiom system that I will take is a modification of System $\mathcal{K}$. It also uses $\sim$, $\wedge$, $\vee$, $\supset$ as four independent primitives, and has the following nine axiom schemes:

S_1: $(X \wedge Y) \supset X$
S_2: $(X \wedge Y) \supset Y$
S_3: $[(X \wedge Y) \supset Z] \supset [X \supset (Y \supset Z)]$ "Exportation, to be abbreviated "Exp."
S_4: $[(X \supset Y) \wedge (X \supset (Y \supset Z))] \supset (X \supset Z)$
S_5: $X \supset (X \vee Y)$
S_6: $Y \supset (X \vee Y)$
S_7: $[(X \supset Z) \wedge (Y \supset Z)] \supset ((X \vee Y) \supset Z)$
S_8: $[(X \supset Y) \wedge (X \supset \sim Y)] \supset \sim X$
S_9: $\sim\sim X \supset X$

I will refer to this system as *System* $\mathcal{S}_0$. As in the other modern systems discussed recently above, I use Modus Ponens as the single rule of inference. The reader can verify that all the axioms are tautologies (i.e. are valid), and Modus Ponens obviously preserves validity, so that the system $\mathcal{S}_0$ is *correct*.

We aim to show that the system $\mathcal{S}_0$ is *complete*, in the sense that all tautologies are provable in the system. There are several ways this can be done. One way is to show how from a truth table for a tautology X, one can find a proof of X in the axiom system $\mathcal{S}_0$. This was the approach I took in my book *Logical Labyrinths* [2009]. What I will do here is essentially to show how from a completed tableau for $\sim X$ we can obtain a proof of X in the axiom system $\mathcal{S}_0$ (but we will pass through an intermediate system along the way). Thus we will establish the completeness of the axiom system as a consequence of the completeness of the tableau method. To do this, we must first prove many things along the way.

In what follows, we abbreviate "X is provable in the system $\mathcal{S}_0$" by "$\vdash X$." We abbreviate "If X is provable in $\mathcal{S}_0$, so is Y" by "$X \vdash Y$". We abbreviate "If X and Y are both provable in $\mathcal{S}_0$, so is Z" by "$X, Y \vdash Z$". A statement of the latter two forms may be said to be new inferences rules that have been proven to be correct.

Problem 0. Prove the following preliminary facts.

F_1. If $\vdash X \supset Y$ then $X \vdash Y$
F_2. If $\vdash X \supset (Y \supset Z)$ then $X, Y \vdash Z$

F_3. $(X \wedge Y) \supset Z \vdash X \supset (Y \supset Z)$ "Exportation," to be abbreviated as "Exp."
F_4. If $\vdash (X \wedge Y) \supset Z$, then $X, Y \vdash Z$.

Problem 1. Using just the axiom schemes S_1, S_2, S_3, S_4, show each of the following:

T_1. $\vdash X \supset (Y \supset Y)$
T_2. $\vdash Y \supset Y$
T_3. $X \supset Y, X \supset (Y \supset Z) \vdash X \supset Z$
T_4. $\vdash X \supset (Y \supset X)$
T_5. $X \vdash Y \supset X$
T_6. $X \supset Y, Y \supset Z \vdash X \supset Z$ "Syllogism," to be abbreviated "Syl."
T_7. $\vdash X \supset (Y \supset (X \wedge Y))$
T_8. $X, Y \vdash X \wedge Y$
T_9. $X \supset Y, Y \supset Z \vdash X \supset (Y \wedge Z)$
T_{10}. $\vdash (X \wedge Y) \supset (Y \wedge X)$
T_{11}. $\vdash (X \wedge (X \supset Y)) \supset Y$
T_{12}. $X \supset (Y \supset Z) \vdash (X \wedge Y) \supset Z$ "Importation," to be abbreviated "Imp."
T_{13}. $X \supset (Y \supset Z) \vdash Y \supset (X \supset Z)$

Recall that we abbreviated $(X_1 \wedge X_2) \wedge X_3$ by $X_1 \wedge X_2 \wedge X_3$

T_{14}. (a) $\vdash (X_1 \wedge X_2 \wedge X_3) \supset X_1$
(b) $\vdash (X_1 \wedge X_2 \wedge X_3) \supset X_2$
(c) $\vdash (X_1 \wedge X_2 \wedge X_3) \supset X_3$
T_{15}. $\vdash ((X \supset Y) \wedge (Y \supset Z)) \supset (X \supset Z)$

Next we wish to establish some results using the schemes S_1, S_2, S_3, S_4, as before, but now also together with S_8 and S_9. These are results about formulas using the three connectives $\supset$, $\wedge$, and $\sim$.

Problem 2. Using just the schemes S_1, S_2, S_3, S_4 and S_8, S_9, prove each of the following:

T_{16}. $(X \wedge Y) \supset Z, (X \wedge Y) \supset \sim Z \vdash X \supset \sim Y$
T_{17}. (a) $\vdash (X \supset Y) \supset (\sim Y \supset \sim X)$
(b) $\vdash (X \supset \sim Y) \supset (Y \supset \sim X)$
(c) $\vdash (\sim X \supset Y) \supset (\sim Y \supset X)$
(d) $\vdash (\sim X \supset \sim Y) \supset (Y \supset X)$
T_{18}. (a) $X \supset Y \vdash \sim Y \supset \sim X$
(b) $X \supset \sim Y \vdash Y \supset \sim X$
(c) $\sim X \supset Y \vdash \sim Y \supset X$
(d) $\sim X \supset \sim Y \vdash Y \supset X$

T_{19}. $\vdash \sim X \supset (X \supset Y)$
T_{20}. (a) $\vdash \sim(X \supset Y) \supset X$
(b) $\vdash \sim(X \supset Y) \supset \sim Y$
T_{21}. $\vdash X \supset \sim\sim X$
T_{22}. $\vdash (X \wedge \sim Y) \supset \sim(X \supset Y)$
T_{23}. (a) $\vdash \sim X \supset \sim(X \wedge Y)$
(b) $\vdash \sim Y \supset \sim(X \wedge Y)$
T_{24}. (a) $X \supset \sim X \vdash \sim X$
(b) $\sim X \supset X \vdash X$

Exercise 1. The two axiom schemes S_8 and S_9 can be replaced by the single axiom: $S_8' : ((\sim X \supset \sim Y) \wedge (\sim X \supset \sim\sim Y)) \supset X$.

That is, from S_8' and all the other schemes except S_8 and S_9, the schemes S_8 and S_9 can both be derived. How? Hint: Successively derive:

(a) S_9
(b) $(X \supset Y) \supset (\sim\sim X \supset Y)$
(c) S_8

Note: There is a certain advantage in using the two schemes S_8 and S_9 in place of the single scheme S_8', in that in the school known as *Intuitionistic Logic*, the schemes S_1 through S_8 are valid, but S_9 is not. Hence having both S_8 and S_9 is helpful in separating intuitionist logic from classical logic, which uses S_9. This separation is skillfully carried out in Kleene's excellent *Introduction to Metamathematics* [1952], in which he puts a special mark on those theorems that use S_9.

Problem 3. Using all the schemes S_1 through S_9, show each of the following:

T_{25}. (a) $\vdash \sim(X \vee Y) \supset \sim X$
(b) $\vdash \sim(X \vee Y) \supset \sim Y$
T_{26}. $X \supset Z, Y \supset Z \vdash (X \vee Y) \supset Z$
T_{27}. $\vdash (\sim X \wedge \sim Y) \supset \sim(X \vee Y)$
T_{28}. $\vdash \sim(X \wedge Y) \supset (\sim X \vee \sim Y)$
T_{29}. $\vdash (X \supset Y) \supset (\sim X \vee Y)$
T_{30}. $\vdash (X \vee Y) \supset (\sim X \supset Y)$
T_{31}. $\vdash \sim(\sim X \wedge \sim Y) \supset (X \vee Y)$
T_{32}. $\vdash (\sim X \supset Y) \supset (X \vee Y)$
T_{33}. $X \vee Y, X \vee Z \vdash X \vee (Y \wedge Z)$
T_{34}. $\vdash X \vee \sim X$
T_{35}. $X \supset Y, \sim X \supset Y \vdash Y$
T_{36}. $\vdash (\sim X \vee Y) \supset (X \supset Y)$
T_{37}. $\sim(X \wedge Y) \vdash X \supset \sim Y$

T_{38}. $X \supset Y, \sim(X \wedge Y) \vdash \sim X$
T_{39}. $X \supset (Y_1 \vee Y_2), \sim(X \wedge Y_1), \sim(X \wedge Y_2) \vdash \sim X$
T_{40}. $Y \supset X, X \vee Z, Z \supset Y \vdash X$
T_{41}. $Y \supset X, X \vee Y_1, X \vee Y_2, (Y_1 \wedge Y_2) \supset Y \vdash X$

T_{42}. Using mathematical induction, show that for all $n \geq 2$ and $i \leq n$, the following holds:

(a) $\vdash (X_1 \wedge X_2 \wedge \cdots \wedge X_n) \supset X_i$
(b) $\vdash X_i \supset (X_1 \vee \cdots \vee X_n)$

Uniform Systems

We now turn to the uniform α, β notation for unsigned formulas introduced in the last chapter. In uniform notation, we have shown that the following are provable in $\mathcal{S}_0$:

Fact A. $(\alpha_1 \wedge \alpha_2) \supset \alpha$, *i.e.*	(1) $(X \wedge Y) \supset (X \wedge Y)$	By T_2.
	(2) $(\sim X \wedge \sim Y) \supset \sim(X \vee Y)$	T_{27}.
	(3) $(X \wedge \sim Y) \supset \sim(X \supset Y)$	T_{22}.
Fact A_1. $\alpha \supset \alpha_1$, *i.e.*	(1) $(X \wedge Y) \supset X$	S_1.
	(2) $\sim(X \vee Y) \supset \sim X$	T_{25} (a).
	(3) $\sim(X \supset Y) \supset X$	T_{20} (a).
	(4) $\sim\sim X \supset X$	S_9.
Fact A_2. $\alpha \supset \alpha_2$, *i.e.*	(1) $(X \wedge Y) \supset Y$	S_2.
	(2) $\sim(X \vee Y) \supset \sim Y$	T_{25} (b).
	(3) $\sim(X \supset Y) \supset \sim Y$	T_{20} (b).
Fact B. $\beta \supset (\beta_1 \vee \beta_2)$, *i.e.*	(1) $(X \vee Y) \supset (X \vee Y)$	By T_2.
	(2) $\sim(X \wedge Y) \supset (\sim X \vee \sim Y)$	T_{28}.
	(3) $(X \supset Y) \supset (\sim X \vee Y)$	T_{29}.
Fact B_1. $\beta_1 \supset \beta$, *i.e.*	(1) $X \supset (X \vee Y)$	S_5.
	(2) $\sim X \supset \sim(X \wedge Y)$	T_{23} (a).
	(3) $\sim X \supset (X \supset Y)$	T_{19}.
Fact B_2. $\beta_2 \supset \beta$, *i.e.*	(1) $Y \supset (X \vee Y)$	S_6.
	(2) $\sim Y \supset \sim(X \wedge Y)$	T_{23} (b).
	(3) $Y \supset (X \supset Y)$	By T_4.

Problem 4. Prove the following four facts:

A_1: $X \supset \alpha, (X \wedge \alpha_1) \supset Y \vdash X \supset Y$.
A_2: $X \supset \alpha, (X \wedge \alpha_2) \supset Y \vdash X \supset Y$.
B: $X \supset \beta, (X \wedge \beta_1) \supset Y, (X \wedge \beta_2) \supset Y \vdash X \supset Y$
C: $X \supset Z, X \supset \sim Z \vdash X \supset Y$
D: $X \supset \sim X \vdash \sim X$

A Uniform System U_1

We recall that an axiom system for Propositional Logic is called *complete* if all tautologies are provable in the system. I shall prove something more general than that our system $\mathcal{S}_0$ is complete – I shall introduce another axiom system U_1 and first show that everything provable in U_1 is also provable in $\mathcal{S}_0$, and then show that the system U_1 is complete. Of course this will show that $\mathcal{S}_0$ is also complete. But note that it also shows that U_1 is *correct*, since no invalid formula could be provable in U_1 because if an invalid formula X were provable in U_1, X would also be provable in $\mathcal{S}_0$, which we have already shown to be correct. It is also easy to prove the correctness of U_1 directly by showing that all its axioms are valid, i.e. are tautologies, and that its rules of inference preserve validity. Note that if we didn't know that $\mathcal{S}_0$ and U_1 were correct, showing that U_1 is complete and that that implies that $\mathcal{S}_0$ is consequently complete might be providing uninteresting information, because, for example, all formulas, including invalid ones, could be provable in both systems.

Note that although the connectives $\sim$ and $\wedge$ are the only logical connectives appearing explicitly in the axioms and rules of U_1 which I am now about to explain to you, the formulas going into θ's can be assumed to have all the connectives used in $\mathcal{S}_0$.[1]

In what follows, we let θ be a finite sequence $\langle X_1, \ldots, X_n \rangle$ of formulas. The formulas $X_1, \ldots, X_n$ are called the *terms* of the sequence. We shall call θ *closed* if the set of its terms contains some formula and its negation. By $C(\theta)$ we shall mean the n-fold conjunction $X_1 \wedge \cdots \wedge X_n$; for $n = 1$, $C(\theta)$ is the single formula X_1. For any formula Y, by (θ, Y) is meant the sequence $\langle X_1, \ldots, X_n, Y \rangle$. We note that $C(\theta, Y) = C(\theta) \wedge Y$.

The system U_1 has just one axiom scheme and four inference rules:

Axioms. All formulas of the form $\sim C(\theta)$, where θ is closed.

Inference Rules

Rules A_1 and A_2: If α is a term of θ, then:

$$A_1: \frac{\sim C(\theta, \alpha_1)}{\sim C(\theta)} \qquad A_2: \frac{\sim C(\theta, \alpha_2)}{\sim C(\theta)}$$

Rule B: If β is a term of θ, then:

$$\frac{\sim C(\theta, \beta_1), \quad \sim C(\theta, \beta_2)}{\sim C(\theta)}$$

[1] The system U_1 actually has the interesting feature of being *uniform*, in the sense that it does not depend on which logical connectives are taken as starters (providing, of course, that all logical connective are definable from them). Although the symbols $\sim$ and $\wedge$ explicitly appear in the presentation of U_1 here, they are not necessarily starters of the system; each or both of them could be defined from the starters.

Rule N: $$\frac{\sim\sim X}{X}$$

Note: *If we let* $X = C(\theta)$, then Rules A_1, A_1, and B can be displayed like this:

A_1: If α is a term of θ, then $$\frac{\sim(X \wedge \alpha_1)}{\sim X}$$

A_2: If α is a term of θ, then $$\frac{\sim(X \wedge \alpha_2)}{\sim X}$$

B: If β is a term of θ, then $$\frac{\sim(X \wedge \beta_1), \quad \sim(X \wedge \beta_2)}{\sim X}$$

We first wish to show that everything provable in U_1 is provable in $\mathcal{S}_0$. Then we will show that all tautologies are provable in U_1, and hence also provable in $\mathcal{S}_0$. Thus we are to establish:

Theorem 1. Everything provable in U_1 is provable in $\mathcal{S}_0$.

Theorem 2. All tautologies are provable in U_1.

To prove Theorem 1, we must first show that all axioms of U_1 are provable in $\mathcal{S}_0$, and then show that each of the inference rules of U_1 holds in $\mathcal{S}_0$, in the sense that if the premises of the rule are provable in $\mathcal{S}_0$, then so is the conclusion.

Problem 5. Prove Theorem 1 by showing:

(a) All axioms of U_1 are provable in $\mathcal{S}_0$.
(b) Each of the inference rules of U_1 holds in $\mathcal{S}_0$.

Next, we wish to prove Theorem 2. We do this by showing how a proof in U_1 of X can be found from a closed tableau for $\sim X$. Theorem 2 will then follow from the completeness of the tableau method, proved in the last chapter: if X is a tautology, then there is a closed tableau for $\sim X$.

To this end, let us temporarily call *a sequence* θ (of formulas) *bad* if $\sim C(\theta)$ is provable in U_1. We identify a formula X with the sequence $\langle X \rangle$ whose only term is X, and $C(X)$ is simply X, and so X is bad if and only if $\sim X$ is provable in U_1. Obviously every closed sequence θ is bad (since $\sim C(\theta)$ is then an axiom of U_1). Rules A_1, A_1, and B can be restated as follows:

A_1: If α is a term of θ and θ, α_1 is bad, so is θ.
A_2: If α is a term of θ and θ, α_2 is bad, so is θ.
B: If β is a term of θ and θ, β_1 and θ, β_2 are both bad, so is θ.

For any tableau $\mathcal{T}$, we call a tableau $\mathcal{T}'$ an *immediate extension* of $\mathcal{T}$ if $\mathcal{T}'$ is obtained from $\mathcal{T}$ by one application of tableau Rule A or tableau Rule B.

We shall call *a tableau $\mathcal{T}$ bad* if every branch of $\mathcal{T}$ is bad.

Problem 6. Suppose $\mathcal{T}'$ is an immediate extension of $\mathcal{T}$. Show that if $\mathcal{T}'$ is bad, so is $\mathcal{T}$.

Problem 7. It then easily follows that if there is a closed tableau $\mathcal{T}$ for X, then X is bad. Why? [Hint: Show that at every stage of the construction of $\mathcal{T}$, the tableau construction at that stage is bad.]

Problem 8. Now complete the proof of Theorem 2.

The reader may remember that I said a little while back that the discussion we were about to embark upon would provide a method of taking a tableau proof of a tautology and creating a proof of the same tautology in U_1 from it. Instead, it may appear that I have given the reader something that many might call a "pure existence proof." And in fact there are many existence proofs that prove the existence of something without saying how to find or construct it; indeed, often without giving any further details about it other than that it exists. This is not the case here, however. For, hidden in the existence proof just given is the method of directly constructing a proof in U_1 of a tautology X from any tableau proof of X using unsigned formulas. I am going to tell you the method now, but I leave it to you to try to figure out how you could have used your understanding of the existence proof just given to come to this method. Following right after my explanation of the method is an example of its use, which the reader might like to consult during the explanation.

So assume we are given a closed tableau (using unsigned formulas) for a formula $\sim X$ (i.e. a proof that the formula X is a tautology). For each branch θ in the closed tableau for $\sim X$, form the formula $\sim C(\theta)$. (Note that in forming $\sim C(\theta)$, we will be assuming we have written down every formula on each branch from left to right in the order from the top of the tableau to the bottom, and have included the duplicates that could arise in cases when a particular formula was concluded more than once on a given branch as a consequence of different tableau rule uses.)

We start our proof in U_1 with this list of formulas, all of which are axioms (because each contains some formula and its negation, due to each branch of the tableau being closed). Then we work backwards through the tableau step by step, mimicking the tableau rules that were applied in constructing the tableau one by one (i.e. starting with the final application of a tableau rule in constructing the tableau and working backwards to the first application of a tableau rule to $\sim X$). What we must guarantee for ourselves, at every stage in our work, is that for every branch $\theta*$ in the tableau as the tableau stood at the end of the i-th application of a tableau rule there will be a corresponding formula $\sim C(\theta^*)$ in the U_1 proof constructed so far. We have guaranteed this to be true at the end the tableau construction, which gave us the first step in the construction of the U_1 proof. After this, we will be adding a single line to the U_1 proof for every application of a tableau rule, and each line will be the result of shortening one or two earlier lines in the U_1 proof, as the tableau rule application directs us.

To be precise, assume we are looking at the tableau as it stood at the end of the j-th step in its construction, and that the partial proof we have constructed in U_1 has a formula $\sim C(\theta)$ as a line in it *for every branch θ in the tableau at this stage of the construction* (along with other lines corresponding to all the other branches that were created later!). If the tableau at this stage was the result of Rule A being used to extend a branch θ^* (a branch in the tableau at the end of step $j - 1$) by either of the two consequences α_1 or α_2, one of the formulas in the U_1 proof constructed so far is $\sim C(\theta^*, \alpha_i)$, $i = 1$ or $i = 2$ (indeed with α_i being the last formula on the branch extended from the branch θ^*, and also the last concatenation component in the formula in the proof corresponding to that branch). But we also know α is one of the formulas in θ^*, so that in our U_1 proof we can conclude $\sim C(\theta^*)$ from $\sim C(\theta^*, \alpha_i)$ by the U_1 Rule A_i. Thus we can add $\sim C(\theta^*)$ to our U_1 proof. The branch θ^* was the only branch that changed at step $j - 1$ of, and now we have added to the U_1 proof the formula corresponding to that rule application at that step. So the U_1 proof now includes formulas for all the branches in the tableau proof at the step $j - 1$. The reasoning when the tableau rule applied was Rule B is similar, except that we must deduce the formula we add to the U_1 proof from two previous formulas in the U_1 proof, namely from the formulas that correspond to the two branches created by the extension of a single branch by a β-rule. So we see that at each stage we are introducing new formulas into the U_1 proof, and that each formula is the same as a formula already in the U_1 proof except that it has one less concatenation. So the new lines in the proof are getting shorter and shorter, and, moreover, we are always eliminating formulas from the concatenations that were introduced into the tableau proof later than the ones remaining. Clearly at some point, the last line in the U_1 proof will be the formula $\sim C(\sim X)$, which is the same as the formula $\sim\sim X$, because it is the only formula in the tableau proof that didn't arrive in the proof as the result of the application of a tableau rule. Recall also that $\sim X$ has been in every concatenation in the U_1 proof, since it was the left-most formula in every axiom placed into the proof at the beginning. But as our last step in the creation of the U_1 proof, we can now immediately conclude X from $\sim\sim X$ by Rule N of U_1.

To illustrate the method, here is a tableau proof (using unsigned formulas) of the tautology $(X \supset Y) \supset (\sim X \vee Y)$. The U_1 proof created from this tableau proof follows immediately afterwards.

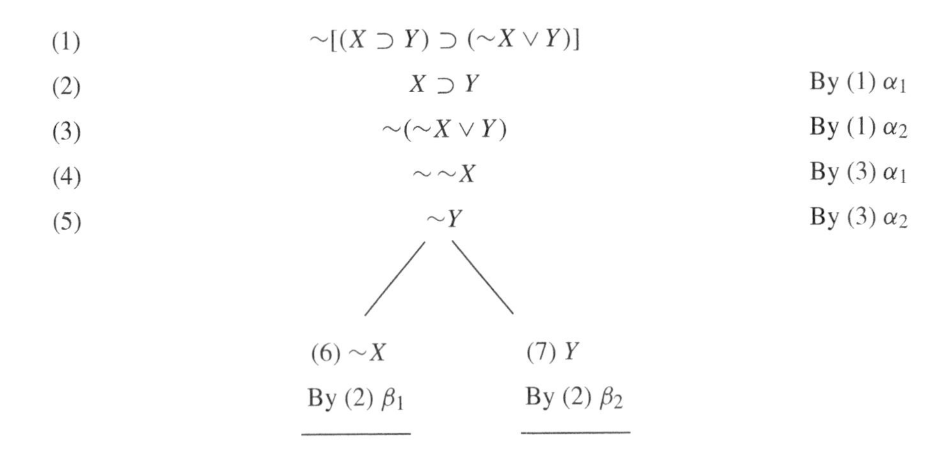

Given this closed tableau (which has its formulas labeled with the order in which they entered the tableau), at the first stage of the construction in the U_1 proof, we form two U_1 axioms from the two branches in our tableau. Then we work backwards through the applications of the tableau rules.

(1) $\sim\{\sim[(X \supset Y) \supset (\sim X \vee Y)] \wedge (X \supset Y) \wedge \sim(\sim X \vee Y) \wedge \sim\sim X \wedge \sim Y \wedge \sim X\}$
Axiom formed from the left-most branch of the tableau

(2) $\sim\{\sim[(X \supset Y) \supset (\sim X \vee Y)] \wedge (X \supset Y) \wedge \sim(\sim X \vee Y) \wedge \sim\sim X \wedge \sim Y \wedge Y\}$
Axiom formed from the right-most branch of the tableau

(3) $\sim\{\sim[(X \supset Y) \supset (\sim X \vee Y)] \wedge (X \supset Y) \wedge \sim(\sim X \vee Y) \wedge \sim\sim X \wedge \sim Y\}$
By (1), (2), Rule B

(4) $\sim\{\sim[(X \supset Y) \supset (\sim X \vee Y)] \wedge (X \supset Y) \wedge \sim(\sim X \vee Y) \wedge \sim\sim X\}$
By (3), Rule A_2

(5) $\sim\{\sim[(X \supset Y) \supset (\sim X \vee Y)] \wedge (X \supset Y) \wedge \sim(\sim X \vee Y)\}$
By (4), Rule A_1

(6) $\sim\{\sim[(X \supset Y) \supset (\sim X \vee Y)] \wedge (X \supset Y)\}$
By (5), Rule A_2

(7) $\sim\{\sim[(X \supset Y) \supset (\sim X \vee Y)]\}$
By (6), Rule A_1

(8) $(X \supset Y) \supset (\sim X \vee Y)$
By (7), Rule N

It can be interesting to compare the proof of this formula using the axioms and rules of the system $\mathcal{S}_0$ with this proof in the axiom system U_1. This tautology was named T_{29} in Problem 3. The proof given in the solution set has eight lines, just like the proof above, but many of those lines involve other tautologies and inferences rules proven valid previously. But even if the U_1 proof is much shorter than the full implicit proof from axioms in the axiom system $\mathcal{S}_0$, it might be difficult to come up with the appropriate U_1 axioms on one's own, just given a formula to prove, for they aren't intuitive, as the axioms of $\mathcal{S}_0$ are. And that would be even more the case if the formula being proved were more complex (i.e. so that a tableau proof would be forced to involve many branchings). I would like to suggest that what this example shows is that the tableau system (with or without signs on the formulas) is not only very intuitive but even often the most efficient way to prove a tautology (not always, though, as one can see by comparing the proofs you have seen in the system $\mathcal{S}_0$ with the most efficient tableau proof you can come up with). But perhaps more importantly, in a tableau proof, one is always doing something very obvious, that is, exploring immediate consequences, which is very easy to do, even if sometimes the best order in which to do so may not be immediately apparent. Or to put it slightly differently, people often say that tautologies are valid due simply to their structure and to the meaning of the logical connectives, and the tableau method allows one to *systematically take the structure of a formula apart* and see what that structure says about all the possible assignments of truth values to the basic components that will make the original formula true. In any case, the principal value of the system U_1 (perhaps the closest axiom system to the tableau system that could be imagined!) is likely to be just that it made the completeness proof of the much more traditional axiom system $\mathcal{S}_0$ very easy, i.e. it was an extremely valuable system for the purpose of examining the *metatheory* of another system.

But it can also be claimed that we have seen how to go from a tableau proof of X to a proof of X in $\mathcal{S}_0$ itself, if we can see how to transform our U_1 proof into a proof in $\mathcal{S}_0$. But in Problem 5 we have seen how to prove all the axioms of U_1 in $\mathcal{S}_0$, and how to prove in $\mathcal{S}_0$ that the inference rules in U_1 preserve validity. The latter fact means that we can actually use the inference rules of U_1 in proofs in $\mathcal{S}_0$. This means our example U_1 proof can be turned into a proof in $\mathcal{S}_0$ simply by citing as the justification of the two axioms it begins with not that they are axioms of U_1, but rather that all axioms of U_1 have been proved in $\mathcal{S}_0$!

Another Uniform System U_2

Here is another uniform system U_2, which in many respects is preferable to the system U_1. Its completeness proof is less roundabout than that of system U_1. It yields an alternative completeness proof for the system $\mathcal{S}_0$ that is more direct. The completeness proof for U_2 is based on the *dual tableaux* of the last chapter. Instead of obtaining a proof X in U_1 from a closed tableau for $\sim X$, we now obtain a proof of X in U_2 from a closed *dual* tableau for X itself. The rule of U_1 that X is inferable from $\sim\sim X$ will now not be necessary. Again there is one axiom scheme for U_2, and instead of the four inference rules for U_1, we will have three inference rules for U_2. Let us begin.

Again, we let θ be a finite sequence $\langle X_1, \ldots, X_n \rangle$ of formulas. By $D(\theta)$ we shall mean the n-fold *disjunction* $X_1 \vee \cdots \vee X_n$. We note that $D(\theta, Y) = D(\theta) \vee Y$. Again we call θ *closed* if its set of terms contains some formula and its negation.

If θ is closed, then unlike the fact that $C(\theta)$ is a contradiction, the formula $D(\theta)$ is a tautology!

Here are the axioms and inference rules of U_2.

Axioms. All formulas $D(\theta)$, where θ is closed.

Rule A^0: If α is a term of θ, then:

$$\frac{D(\theta, \alpha_1), \ D(\theta, \alpha_2)}{D(\theta)}$$

Rules B^0: If β is a term of θ, then:

$$B_1{}^0 : \quad \frac{D(\theta, \beta_1)}{D(\theta)} \qquad\qquad B_2{}^0 : \quad \frac{D(\theta, \beta_2)}{D(\theta)}$$

This concludes the description of the system U_2.

We shall call θ *nice* if $D(\theta)$ is provable in U_2.

Thus the axioms and inference rules of U_2 can be restated thus:

Axioms. All closed sequences θ are nice.

Rules:

A^0: If α is a term of θ and θ, α_1 is nice, and θ, α_2 is nice, so is θ.
$B_1{}^0$: If β is a term of θ and θ, β_1 is nice, so is θ.
$B_2{}^0$: If β is a term of θ and θ, β_2 is nice, so is θ.

Theorem 1^0. All provable formulas of U_2 are provable in $\mathcal{S}_0$.

Theorem 2^0. All tautologies are provable in U_2.

Problem 9. Prove Theorem 1^0.

We call a dual tableau $\mathcal{T}'$ an immediate extension of a dual tableau $\mathcal{T}$ if $\mathcal{T}'$ is obtained from $\mathcal{T}$ by one application of dual tableau Rule A^0 or Rule B^0.

We call a dual tableau $\mathcal{T}$ *nice* if every branch of $\mathcal{T}$ is nice.

Problem 10. (a) First show that if a dual tableau $\mathcal{T}'$ is an immediate extension of the dual tableau $\mathcal{T}$, then, if $\mathcal{T}'$ is nice, so is $\mathcal{T}$.

(b) Next, consider a sequence $\mathcal{T}_0, \mathcal{T}_1, \ldots, \mathcal{T}_n$, of dual tableaux such that for each $i \leq n$, the dual tableau $\mathcal{T}_{i+1}$ is an immediate extension of $\mathcal{T}_i$. Show that if $\mathcal{T}_n$ is closed, then all the dual tableaux $\mathcal{T}_0, \mathcal{T}_1, \ldots, \mathcal{T}_n$, are nice.

(c) Finally, complete the proof of Theorem 2^0.

The correctness of U_2 also follows as a consequence from the correctness of $\mathcal{S}_0$ and from the fact that all provable formulas of U_2 are provable in $\mathcal{S}_0$. It can be proved very easily directly as well.

I would like to briefly mention another uniform system that is essentially the one I introduced in my book *Logical Labyrinths* [Smullyan, 2009]. The axioms of the system are all formulas of the form $(X_1 \wedge \cdots \wedge X_n) \supset X_i$ $(i \leq n)$.

The inference rules are the following:

I. $$\frac{X \supset \alpha_1,\ X \supset \alpha_2}{X \supset \alpha}$$

II. (a) $$\frac{X \supset \beta_1}{X \supset \beta}$$ (b) $$\frac{X \supset \beta_2}{X \supset \beta}$$

III. $$\frac{(X \wedge Y) \supset Z,\ (X \wedge {\sim}Y) \supset Z}{X \supset Z}$$

IV. $$\frac{X \supset Z,\ {\sim}X \supset Z}{Z}$$

Note: In the system of [Smullyan, 2004], instead of III, I had the exportation rule:

$$\frac{(X \wedge Y) \supset Z}{X \supset (Y \supset Z)}$$

I obtained the completeness of the system by showing how a truth table for a tautology yields a proof of the tautology in this system. The ambitious reader might like to try this as an exercise. The curious reader can refer to [Smullyan, 2004].

Solutions to the Problems of Chapter 7

Note: Modus Ponens will be abbreviated M.P. here.

0. Solutions to the 4 parts:

F_1. To show: If $\vdash X \supset Y$ then $X \vdash Y$.

(1) $\vdash X \supset Y$ — Hypothesis.
(2) $\vdash X$ — Hypothesis.
(3) $\vdash Y$ — By (2), (1), M.P.

Thus $\vdash X \supset Y$ implies that $X \vdash Y$.

F_2. To show: If $\vdash X \supset (Y \supset Z)$ then $X, Y \vdash Z$.

(1) $\vdash X \supset (Y \supset Z)$ — Hypothesis.

We are to show that then $X, Y \vdash Z$, i.e. that $\vdash X$ and $\vdash Y$ implies $\vdash Z$. Hence suppose (2) and (3) below

(2) $\vdash X$
(3) $\vdash Y$ — Then
(4) $\vdash Y \supset Z$ — By (2), (1), M.P.
(5) $\vdash Z$ — By (3), (4), M.P.

Thus $\vdash X$ and $\vdash Y$ implies $\vdash Z$ (under hypothesis 1). Thus hypothesis (1) implies $X, Y \vdash Z$.

F_3. To show: $(X \wedge Y) \supset Z \vdash X \supset (Y \supset Z)$.

(1) $\vdash (X \wedge Y) \supset Z$ — Hypothesis.

(2) $\vdash [(X \wedge Y) \supset Z] \supset [X \supset (Y \supset Z)]$ — S_3.

(3) $\vdash X \supset (Y \supset Z)$ — By (1), (2), M.P.

Thus, $(X \wedge Y) \supset Z$ implies $\vdash X \supset (Y \supset Z)$

F_4. To show: If $\vdash (X \wedge Y) \supset Z$, then $X, Y \vdash Z$.

(1) $\vdash (X \wedge Y) \supset Z$ — Hypothesis.

(2) $\vdash X \supset (Y \supset Z)$ — By (1), F_3. (Exp.)

(3) $X, Y \vdash Z$ — By (2), F_2.

Thus, $\vdash (X \wedge Y) \supset Z$ implies $X, Y \vdash Z$.

1. Solutions to the 15 parts:

T_1. To show: $\vdash X \supset (Y \supset Y)$.

(1) $\vdash (X \wedge Y) \supset Y \vdash X \supset (Y \supset Y)$ — By F_3, taking Y for Z.

(2) $\vdash (X \wedge Y) \supset Y$ — S_2.

(3) $\vdash X \supset (Y \supset Y)$ — By (2), (1), M.P.

T_2. To show: $\vdash Y \supset Y$.

(1) $X \supset (Y \supset Y)$ — T_1.

Now take X to be any provable formula, such as an axiom. From this X and (1), $Y \supset Y$ follows by M.P.

T_3. To show: $X \supset Y, X \supset (Y \supset Z) \vdash X \supset Z$

(1) $\vdash X \supset Y$ — Hypothesis.

(2) $\vdash X \supset (Y \supset Z)$ — Hypothesis.

(3) $\vdash [(X \supset Y) \wedge (X \supset (Y \supset Z))] \supset (X \supset Z)$ — S_4.

(4) $\vdash (X \supset Y) \vdash (X \supset (Y \supset Z)) \supset (X \supset Z)$ — By (3), F_3. (Exp.)

(5) $\vdash (X \supset (Y \supset Z)) \supset (X \supset Z)$ — By (1), (4), M.P.

(6) $\vdash X \supset Z$ — By (2), (5), M.P.

T_4. To show: $\vdash X \supset (Y \supset X)$.

(1) $\vdash (X \wedge Y) \supset X$ — S_1.

(2) $\vdash X \supset (Y \supset X)$ — By (1), F_3. (Exp.)

T_5. To show: $X \vdash Y \supset X$.

(1) $\vdash X \supset (Y \supset X)$. — T_4.

(2) $X \vdash Y \supset X$ — By (1), F_1.

T_6. To show: $X \supset Y, Y \supset Z \vdash X \supset Z$. — "Syllogism"

(1) $\vdash X \supset Y$ — Hypothesis.

(2) $\vdash Y \supset Z$ — Hypothesis.

(3) $\vdash X \supset (Y \supset Z)$ By (2), T_5.

(4) $\vdash X \supset Z$ By (1), (3), T_3.

T_7. To show: $\vdash X \supset (Y \supset (X \wedge Y))$

(1) $\vdash (X \wedge Y) \supset (X \wedge Y)$ By T_2.

(2) $\vdash X \supset (Y \supset (X \wedge Y))$ By (1), F_3. (Exp.)

T_8. To show: $X, Y \vdash X \wedge Y$.

(1) $\vdash X \supset (Y \supset (X \wedge Y))$ T_7.

(2) $X, Y \vdash X \wedge Y$ By F_2

T_9. To show: $X \supset Y, X \supset Z \vdash X \supset (Y \wedge Z)$.

(1) $\vdash X \supset Y$ Hypothesis.

(2) $\vdash X \supset Z$ Hypothesis.

(3) $\vdash Y \supset (Z \supset (Y \wedge Z))$ By T_7.

(4) $\vdash X \supset (Z \supset (Y \wedge Z))$ By (1), (3), T_6. (Syl.)

(5) $\vdash X \supset (Y \wedge Z)$ By (2), (4), T_3.

T_{10}. To show: $\vdash (X \wedge Y) \supset (Y \wedge X)$.

(1) $\vdash (X \wedge Y) \supset Y$ S_2.

(2) $\vdash (X \wedge Y) \supset X$ S_1.

(3) $\vdash (X \wedge Y) \supset (Y \wedge X)$ By (1), (2), T_9.

T_{11}. To show: $\vdash (X \wedge (X \supset Y)) \supset Y$.

(1) $\vdash (X \wedge (X \supset Y)) \supset X$ By S_1.

(2) $\vdash (X \wedge (X \supset Y)) \supset (X \supset Y)$ By S_2.

(3) $\vdash (X \wedge (X \supset Y)) \supset Y$ By (1), (2), T_3.

T_{12}. To show: $X \supset (Y \supset Z) \vdash (X \wedge Y) \supset Z$. "Importation", abbreviated "Imp.".

(1) $\vdash X \supset (Y \supset X)$ Hypothesis.

(2) $\vdash (X \wedge Y) \supset X$ S_1.

(3) $\vdash (X \wedge Y) \supset (Y \supset Z)$ By (2), (1), T_6. (Syl.)

(4) $\vdash (X \wedge Y) \supset Y$ S_2.

(5) $\vdash (X \wedge Y) \supset Z$ By (4), (3), T_3.

T_{13}. To show: $X \supset (Y \supset Z) \vdash Y \supset (X \supset Z)$.

(1) $\vdash X \supset (Y \supset Z)$ Hypothesis.

(2) $\vdash (X \wedge Y) \supset Z$ By (1), T_{12}.

(3) $\vdash (Y \wedge X) \supset (X \wedge Y)$ By T_{10}.

(4) $\vdash (Y \wedge X) \supset Z$ By (3), (2), T_6. (Syl.)

(5) $\vdash Y \supset (X \supset Z)$ By (4), F_3. (Exp.)

T_{14}. To show: (a) $\vdash (X_1 \wedge X_2 \wedge X_3) \supset X_1$.
(b) $\vdash (X_1 \wedge X_2 \wedge X_3) \supset X_2$.
(c) $\vdash (X_1 \wedge X_2 \wedge X_3) \supset X_3$.

(1) $\vdash ((X \wedge Y) \wedge Z) \supset (X \wedge Y)$	By S_1.	
(2) $\vdash (X \wedge Y) \supset X$	S_1.	
(3) $\vdash (X \wedge Y) \supset Y$	S_2.	
(4) $\vdash ((X \wedge Y) \wedge Z) \supset X$	By (1), (2), T_6. (Syl.)	(a)
(5) $\vdash ((X \wedge Y) \wedge Z) \supset Y$	By (1), (3), T_6. (Syl.)	(b)
(6) $\vdash ((X \wedge Y) \wedge Z) \supset Z$	By S_2.	(c)

T_{15}. To show: $\vdash ((X \supset Y) \wedge (Y \supset Z)) \supset (X \supset Z)$.

(1) $\vdash ((X \supset Y) \wedge (Y \supset Z) \wedge X) \supset (X \supset Y)$	By T_{14}.
(2) $\vdash ((X \supset Y) \wedge (Y \supset Z) \wedge X) \supset (Y \supset Z)$	By T_{14}.
(3) $\vdash ((X \supset Y) \wedge (Y \supset Z) \wedge X) \supset X$	By T_{14}.
(4) $\vdash ((X \supset Y) \wedge (Y \supset Z) \wedge X) \supset Y$	By (3), (1), T_3.
(5) $\vdash ((X \supset Y) \wedge (Y \supset Z) \wedge X) \supset Z$	By (2), (4), T_3.
(6) $\vdash ((X \supset Y) \wedge (Y \supset Z)) \supset (X \supset Z)$	By (5), F_3. (Exp.)

2. Solutions of the 9 parts:

T_{16}. To show: $(X \wedge Y) \supset Z, (X \wedge Y) \supset \sim Z \vdash X \supset \sim Y$.

(1) $\vdash (X \wedge Y) \supset Z$	Hypothesis.
(2) $\vdash (X \wedge Y) \supset \sim Z$	Hypothesis.
(3) $\vdash X \supset (Y \supset Z)$	By (1), F_3. (Exp.)
(4) $\vdash X \supset (Y \supset \sim Z)$	By (2), F_3. (Exp.)
(5) $\vdash X \supset [(Y \supset Z) \wedge (Y \supset \sim Z)]$	By (3), (4), T_9.
(6) $\vdash [(Y \supset Z) \wedge (Y \supset \sim Z)] \supset \sim Y$	By S_8.
(7) $\vdash X \supset \sim Y$	By (5), (6), T_6. (Syl.)

T_{17}.

(a) To show: (a) $\vdash (X \supset Y) \supset (\sim Y \supset \sim X)$.

(1) $\vdash ((X \supset Y) \wedge \sim Y \wedge X) \supset (X \supset Y)$	By T_{14}.
(2) $\vdash ((X \supset Y) \wedge \sim Y \wedge X) \supset \sim Y$	By T_{14}.
(3) $\vdash ((X \supset Y) \wedge \sim Y \wedge X) \supset X$	By T_{14}.
(4) $\vdash ((X \supset Y) \wedge \sim Y \wedge X) \supset Y$	By (3), (1), T_3.
(5) $\vdash ((X \supset Y) \wedge \sim Y) \supset \sim X$	By (4), (2), T_{16}.
(6) $\vdash (X \supset Y) \supset (\sim Y \supset \sim X)$	By (5), F_3. (Exp.)

(b) To show: (b) $\vdash (X \supset \sim Y) \supset (Y \supset \sim X)$.

(1) $\vdash ((X \supset \sim Y) \wedge Y \wedge X) \supset (X \supset \sim Y)$ By T_{14}.

(2) $\vdash ((X \supset \sim Y) \wedge Y \wedge X) \supset Y$ By T_{14}.

(3) $\vdash ((X \supset \sim Y) \wedge Y \wedge X) \supset X$ By T_{14}.

(4) $\vdash ((X \supset \sim Y) \wedge Y \wedge X) \supset \sim Y$ By (3), (1), T_3.

(5) $\vdash ((X \supset \sim Y) \wedge Y) \supset \sim X$ By (4), (2), T_{16}.

(6) $\vdash (X \supset \sim Y) \supset (Y \supset \sim X)$ By (5), F_3. (Exp.)

(c) To show: (c) $\vdash (\sim X \supset Y) \supset (\sim Y \supset X)$.

(1) $\vdash ((\sim X \supset Y) \wedge \sim Y \wedge \sim X) \supset (\sim X \supset Y)$ By T_{14}.

(2) $\vdash ((\sim X \supset Y) \wedge \sim Y \wedge \sim X) \supset \sim Y$ By T_{14}.

(3) $\vdash ((\sim X \supset Y) \wedge \sim Y \wedge \sim X) \supset \sim X$ By T_{14}.

(4) $\vdash ((\sim X \supset Y) \wedge \sim Y \wedge \sim X) \supset Y$ By (3), (1), T_3.

(5) $\vdash ((\sim X \supset Y) \wedge \sim Y) \supset \sim\sim X$ By (2), (4), T_{16}.

(6) $\vdash \sim\sim X \supset X$ S_9.

(7) $\vdash ((\sim X \supset Y) \wedge \sim Y) \supset X$ By (5), (6), T_6. (Syl.)

(8) $\vdash (\sim X \supset Y) \supset (\sim Y \supset X)$ By (7), F_3. (Exp.)

(d) To show: (d) $\vdash (\sim X \supset \sim Y) \supset (Y \supset X)$.

(1) $\vdash ((\sim X \supset \sim Y) \wedge Y \wedge \sim X) \supset (\sim X \supset \sim Y)$ By T_{14}.

(2) $\vdash ((\sim X \supset \sim Y) \wedge Y \wedge \sim X) \supset Y$ By T_{14}.

(3) $\vdash ((\sim X \supset \sim Y) \wedge Y \wedge \sim X) \supset \sim X$ By T_{14}.

(4) $\vdash ((\sim X \supset \sim Y) \wedge Y \wedge \sim X) \supset \sim Y$ By (3), (1), T_3.

(5) $\vdash ((\sim X \supset \sim Y) \wedge Y) \supset \sim\sim X$ By (2), (4), T_{16}.

(6) $\vdash \sim\sim X \supset X$ S_9.

(7) $\vdash ((\sim X \supset \sim Y) \wedge Y) \supset X$ By (5), (6), T_6. (Syl.)

(8) $\vdash (\sim X \supset \sim Y) \supset (Y \supset X)$ By (7), F_3. (Exp.)

T_{18}. To show:

(a) $X \supset Y \vdash \sim Y \supset \sim X$.

(b) $X \supset \sim Y \vdash Y \supset \sim X$.

(c) $\sim X \supset Y \vdash \sim Y \supset X$.

(d) $\sim X \supset \sim Y \vdash Y \supset X$.

These are immediate from T_{17} and F_1.

T_{19}. To show: $\vdash \sim X \supset (X \supset Y)$.

(1) $\vdash((\sim X \wedge X) \wedge \sim Y) \supset X$	By T_{14}.
(2) $\vdash((\sim X \wedge X) \wedge \sim Y) \supset \sim X$	By T_{14}.
(3) $\vdash(\sim X \wedge X) \supset \sim\sim Y$	By (1), (2), T_{16}.
(4) $\vdash \sim\sim Y \supset Y$	By S_9.
(5) $\vdash(\sim X \wedge X) \supset Y$	By (3), (4), T_6. (Syl.)
(6) $\vdash \sim X \supset (X \supset Y)$	By (5), F_3. (Exp.)

T_{20}. (a) To show: $\sim(X \supset Y) \supset X$.

(1) $\vdash \sim X \supset (X \supset Y)$	T_{19}.
(2) $\vdash \sim(X \supset Y) \supset \sim\sim X$	By (1), T_{18} (a).
(3) $\vdash \sim\sim X \supset X$	S_9.
(4) $\vdash \sim(X \supset Y) \supset X$	By (2), (3), M.P.

(b) To show: $\sim(X \supset Y) \supset \sim Y$.

(1) $\vdash Y \supset (X \supset Y)$	By T_4.
(2) $\vdash \sim(X \supset Y) \supset \sim Y$	By (1), T_{18} (a).

T_{21}. To show: $\vdash X \supset \sim\sim X$

(1) $\vdash(X \wedge \sim X) \supset X$	By S_1.
(2) $\vdash(X \wedge \sim X) \supset \sim X$	By S_2.
(3) $\vdash X \supset \sim\sim X$	By (1), (2), T_{16}.

T_{22}. To show: $\vdash(X \wedge \sim Y) \supset \sim(X \supset Y)$.

(1) $\vdash(X \wedge \sim Y \wedge (X \supset Y)) \supset X$	By T_{14}.
(2) $\vdash(X \wedge \sim Y \wedge (X \supset Y)) \supset \sim Y$	By T_{14}.
(3) $\vdash(X \wedge \sim Y \wedge (X \supset Y)) \supset (X \supset Y)$	By T_{14}.
(4) $\vdash(X \wedge \sim Y \wedge (X \supset Y)) \supset Y$	By (1), (3), T_3.
(5) $\vdash(X \wedge \sim Y) \supset \sim(X \supset Y)$	By (2), (4), T_{16}.

T_{23}. (a) To show: $\vdash \sim X \supset \sim(X \wedge Y)$.

(1) $\vdash(X \wedge Y) \supset X$	S_1.
(2) $\vdash \sim X \supset \sim(X \wedge Y)$	By (1), T_{18} (a).

(b) To show: $\vdash \sim Y \supset \sim(X \wedge Y)$.

(1) $\vdash(X \wedge Y) \supset Y$	S_2.
(2) $\vdash \sim Y \supset \sim(X \wedge Y)$	By (1), T_{18} (a).

T_{24}. (a) To show: $X \supset \sim X \vdash \sim X$.

(1) $\vdash X \supset \sim X$	Hypothesis.
(2) $\vdash X \supset X$	By T_2.
(3) $\vdash (X \supset X) \wedge (X \supset \sim X)$	By (1), (2), T_8.
(4) $\vdash [(X \supset X) \wedge (X \supset \sim X)] \supset \sim X$	By S_8.
(5) $\vdash \sim X$	By (3), (4), M.P.

(b) To show: $\sim X \supset X \vdash X$.

(1) $\vdash \sim X \supset X$	Hypothesis.
(2) $\vdash X \supset \sim\sim X$	T_{21}.
(3) $\vdash \sim X \supset \sim\sim X$	By (1), (2), T_6. (Syl.)
(4) $\vdash \sim\sim X$	By (3), 24(a).
(5) $\vdash \sim\sim X \supset X$	S_9.
(6) $\vdash X$	By (2), (3), M.P.

Exercise 1.

(a) To show S_9, i.e. $\sim\sim X \supset X$

(1) $\vdash ((\sim X \supset \sim X) \wedge (\sim X \supset \sim\sim X)) \supset X$	By S_8', taking X for Y.
(2) $\vdash (\sim X \supset \sim X) \supset ((\sim X \supset \sim\sim X) \supset X)$	By (1), F_3. (Exp.)
(3) $\vdash \sim X \supset \sim X$	By T_2.
(4) $\vdash (\sim X \supset \sim\sim X) \supset X$	By (2), (3), $M.P.$
(5) $\vdash \sim\sim X \supset (\sim X \supset \sim\sim X)$	By T_4.
(6) $\vdash \sim\sim X \supset X$	By (4), (5), T_6. (Syl.)

(b) To show: $(X \supset Y) \supset (\sim\sim X \supset Y)$

(1) $\vdash ((\sim\sim X \supset X) \wedge (X \supset Y)) \supset (\sim\sim X \supset Y)$	By T_{15}.
(2) $\vdash (\sim\sim X \supset X) \supset ((X \supset Y) \supset (\sim\sim X \supset Y))$	By (1), F_3. (Exp.)
(3) $\vdash \sim\sim X \supset X$	By Exercise 1(a).
(4) $\vdash (X \supset Y) \supset (\sim\sim X \supset Y)$	By (2), (3) M.P.

(c) To show S_8, i.e. $[(X \supset Y) \wedge (X \supset \sim Y)] \supset \sim X$

(1) $\vdash ((\sim\sim X \supset Y) \wedge (\sim\sim X \supset \sim Y)) \supset \sim X$	By S_8', taking $\sim X$ for X.
(2) $\vdash (\sim\sim X \supset Y) \supset ((\sim\sim X \supset \sim Y) \supset \sim X)$	By (1), F_3. (Exp.)
(3) $\vdash (X \supset Y) \supset (\sim\sim X \supset Y)$	By Exercise 1 (b).
(4) $\vdash (X \supset Y) \supset ((\sim\sim X \supset \sim Y) \supset \sim X)$	By (2), (3), T_6. (Syl.)

(5) $\vdash(\sim\sim X \supset \sim Y) \supset ((X \supset Y) \supset \sim X)$ — By (4), T_{13}.

(6) $\vdash(X \supset \sim Y) \supset (\sim\sim X \supset \sim Y)$ — By Exercise 1 (b).

(7) $\vdash(X \supset \sim Y) \supset ((X \supset Y) \supset \sim X)$ — By (5), (6), T_6. (Syl.)

(8) $\vdash(X \supset Y) \supset ((X \supset \sim Y) \supset \sim X)$ — By (7), T_{13}.

(9) $\vdash((X \supset Y) \wedge (X \supset \sim Y)) \supset \sim X$ — By (8), T_{12}.

3. T_{25}. (a) To show: $\vdash \sim(X \vee Y) \supset \sim X$

(1) $\vdash X \supset (X \vee Y)$ — S_5.

(2) $\sim(X \vee Y) \supset \sim X$ — By (1), T_{18} (a).

(b) To show: $\vdash \sim(X \vee Y) \supset \sim Y$

(1) $\vdash Y \supset (X \vee Y)$ — By S_6.

(2) $\vdash \sim(X \vee Y) \supset \sim Y$ — By (1), T_{18} (a).

T_{26}. To show: $X \supset Z, Y \supset Z \vdash (X \vee Y) \supset Z$

(1) $\vdash[(X \supset Z) \wedge (Y \supset Z)] \supset ((X \vee Y) \supset Z)$ — S_7.

(2) $X \supset Z, Y \supset Z \vdash (X \vee Y) \supset Z$ — By (1), F_4.

T_{27}. To show: $\vdash(\sim X \wedge \sim Y) \supset \sim(X \vee Y)$

(1) $\vdash(\sim X \wedge \sim Y) \supset \sim X$ — By S_1.

(2) $\vdash X \supset \sim(\sim X \wedge \sim Y)$ — By (1), T_{18} (b).

(3) $\vdash(\sim X \wedge \sim Y) \supset \sim Y$ — By S_2.

(4) $\vdash Y \supset \sim(\sim X \wedge \sim Y)$ — By (3), T_{18} (b).

(5) $\vdash(X \vee Y) \supset \sim(\sim X \wedge \sim Y)$ — By (2), (4), T_{26}.

(6) $\vdash(\sim X \wedge \sim Y) \supset \sim(X \vee Y)$ — By (5), T_{18} (b).

T_{28}. To show: $\vdash \sim(X \wedge Y) \supset (\sim X \vee \sim Y)$

(1) $\vdash \sim X \supset (\sim X \vee \sim Y)$ — By S_5.

(2) $\vdash \sim Y \supset (\sim X \vee \sim Y)$ — By S_6.

(3) $\vdash \sim(\sim X \vee \sim Y) \supset X$ — By (1), T_{18} (c).

(4) $\vdash \sim(\sim X \vee \sim Y) \supset Y$ — By (2), T_{18} (c).

(5) $\vdash \sim(\sim X \vee \sim Y) \supset (X \wedge Y)$ — By (3), (4), T_9.

(6) $\vdash \sim(X \wedge Y) \supset (\sim X \vee \sim Y)$ — By (5), T_{18} (c).

T_{29}. To show: $\vdash(X \supset Y) \supset (\sim X \vee Y)$

(1) $\vdash \sim X \supset (\sim X \vee Y)$ — By S_5.

(2) $\vdash Y \supset (\sim X \vee Y)$ — By S_6.

(3) $\vdash \sim(\sim X \vee Y) \supset X$ By (1), T_{18} (c).

(4) $\vdash \sim(\sim X \vee Y) \supset \sim Y$ By (2), T_{18} (a).

(5) $\vdash \sim(\sim X \vee Y) \supset (X \wedge \sim Y)$ By (3), (4), T_9.

(6) $\vdash (X \wedge \sim Y) \supset \sim(X \supset Y)$ By T_{22}.

(7) $\vdash \sim(\sim X \vee Y) \supset \sim(X \supset Y)$ By (5), (6), T_6. (Syl.)

(8) $\vdash (X \supset Y) \supset (\sim X \vee Y)$ By (7), T_{18} (d).

T_{30}. To show: $\vdash (X \vee Y) \supset (\sim X \supset Y)$

(1) $\vdash \sim X \supset (X \supset Y)$ T_{19}.

(2) $\vdash X \supset (\sim X \supset Y)$ By (1), T_{13}.

(3) $\vdash Y \supset (\sim X \supset Y)$ By T_4.

(4) $\vdash (X \vee Y) \supset (\sim X \supset Y)$ By (2), (3), T_{26}.

T_{31}. To show: $\vdash \sim(\sim X \wedge \sim Y) \supset (X \vee Y)$

(1) $\vdash \sim(X \vee Y) \supset \sim X$ By T_{25} (a).

(2) $\vdash \sim(X \vee Y) \supset \sim Y$ By T_{25} (b).

(3) $\vdash \sim(X \vee Y) \supset (\sim X \wedge \sim Y)$ By (1), (2), T_9.

(4) $\vdash \sim(\sim X \wedge \sim Y) \supset (X \vee Y)$ By (3), T_{18} (c).

T_{32}. To show: $\vdash (\sim X \supset Y) \supset (X \vee Y)$

(1) $\vdash (\sim X \wedge \sim Y) \supset \sim(\sim X \supset Y)$ By T_{22} (taking $\sim X$ for X).

(2) $\vdash (\sim X \supset Y) \supset \sim(\sim X \wedge \sim Y)$ By (1), T_{18} (b).

(3) $\vdash \sim(\sim X \wedge \sim Y) \supset (X \vee Y)$ T_{31}.

(4) $\vdash (\sim X \supset Y) \supset (X \vee Y)$ By (2), (3), T_6. (Syl.)

T_{33}. To show: $X \vee Y, X \vee Z \vdash X \vee (Y \wedge Z)$

(1) $\vdash X \vee Y$ Hypothesis.

(2) $\vdash X \vee Z$ Hypothesis.

(3) $\vdash (X \vee Y) \supset (\sim X \supset Y)$ T_{30}.

(4) $\vdash \sim X \supset Y$ By (1), (3), M.P.

(5) $\vdash (X \vee Z) \supset (\sim X \supset Z)$ By T_{30}.

(6) $\vdash \sim X \supset Z$ By (2), (5), M.P.

(7) $\vdash \sim X \supset (Y \wedge Z)$ By (4), (6), T_9.

(8) $\vdash (\sim X \supset (Y \wedge Z)) \supset (X \vee (Y \wedge Z))$ By T_{32}.

(9) $\vdash X \vee (Y \wedge Z)$ By (7), (8), M.P.

T_{34}. To show: $\vdash X \vee \sim X$

(1) $\vdash \sim X \supset \sim X$ By T_2.

(2) $\vdash (\sim X \supset \sim X) \supset (X \vee \sim X)$ By T_{32}.

(2) $\vdash X \vee \sim X$ By (1), (2), M.P.

T_{35}. To show: $X \supset Y,\ \sim X \supset Y \vdash Y$

(1) $\vdash X \supset Y$ Hypothesis.

(2) $\vdash \sim X \supset Y$ Hypothesis.

(3) $\vdash (X \vee \sim X) \supset Y$ By (1), (2), T_{26}.

(4) $\vdash (X \vee \sim X)$ T_{34}.

(5) $\vdash Y$ By (4), (3), M.P.

T_{36}. To show: $\vdash (\sim X \vee Y) \supset (X \supset Y)$

(1) $\vdash \sim X \supset (X \supset Y)$ T_{19}.

(2) $\vdash Y \supset (X \supset Y)$ By T_4.

(3) $\vdash (\sim X \vee Y) \supset (X \supset Y)$ By (1), (2), T_{26}.

T_{37}. To show: $\sim(X \wedge Y) \vdash X \supset \sim Y$

(1) $\vdash \sim(X \wedge Y)$ Hypothesis.

(2) $\vdash \sim(X \wedge Y) \supset (\sim X \vee \sim Y)$ T_{28}.

(3) $\vdash \sim X \vee \sim Y$ By (1), (2), M.P.

(4) $\vdash (\sim X \vee \sim Y) \supset (X \supset \sim Y)$ By T_{36}.

(5) $\vdash X \supset \sim Y$ By (3), (4), M.P.

T_{38}. To show: $X \supset Y,\ \sim(X \wedge Y) \vdash \sim X$

(1) $\vdash X \supset Y$ Hypothesis.

(2) $\vdash \sim(X \wedge Y)$ Hypothesis.

(3) $\vdash X \supset X$ By T_2.

(4) $\vdash X \supset (X \wedge Y)$ By (3), (1), T_9.

(5) $\vdash \sim(X \wedge Y) \supset \sim X$ By (4), T_{17} (a).

(6) $\vdash \sim X$ By (2), (5), M.P.

T_{39}. To show: $X \supset (Y_1 \vee Y_2),\ \sim(X \wedge Y_1),\ \sim(X \wedge Y_2) \vdash \sim X$

(1) $\vdash X \supset (Y_1 \vee Y_2)$ Hypothesis.

(2) $\vdash \sim(X \wedge Y_1)$ Hypothesis.

(3) $\vdash \sim(X \wedge Y_2)$ Hypothesis.

(4) $\vdash X \supset \sim Y_1$ By (2), T_{37}.

(5) $\vdash X \supset \sim Y_2$	By (3), T_{37}.
(6) $\vdash Y_1 \supset \sim X$	By (4), T_{17} (b).
(7) $\vdash Y_2 \supset \sim X$	By (5), T_{17} (b).
(8) $\vdash (Y_1 \vee Y_2) \supset \sim X$	By (6), (7), T_{26}.
(9) $\vdash X \supset \sim X$	By (1), (8), T_6. (Syl.)
(10) $\vdash \sim X$	By (9), T_{24} (a).

T_{40}. $Y \supset X,\ X \vee Z,\ Z \supset Y \vdash X$

(1) $\vdash Y \supset X$	Hypothesis.
(2) $\vdash X \vee Z$	Hypothesis.
(3) $\vdash Z \supset Y$	Hypothesis.
(4) $\vdash (X \vee Z) \supset (\sim X \supset Z)$	By T_{30}.
(5) $\vdash \sim X \supset Z$	By (2), (4), M.P.
(6) $\vdash \sim X \supset Y$	By (5), (3), T_6. (Syl.)
(7) $\vdash \sim X \supset X$	By (6), (1), T_6. (Syl.)
(8) $\vdash X$	By (7), T_{24} (b).

T_{41}. To show: $Y \supset X,\ X \vee Y_1,\ X \vee Y_2,\ (Y_1 \wedge Y_2) \supset Y \vdash X$

(1) $\vdash Y \supset X$	Hypothesis.
(2) $\vdash X \vee Y_1$	Hypothesis.
(3) $\vdash X \vee Y_2$	Hypothesis.
(4) $\vdash (Y_1 \wedge Y_2) \supset Y$	Hypothesis.
(5) $\vdash X \vee (Y_1 \wedge Y_2)$	By (2), (3), T_{33}.
(6) $\vdash (X \vee (Y_1 \wedge Y_2)) \supset (\sim X \supset (Y_1 \wedge Y_2))$	By T_{30}.
(7) $\vdash \sim X \supset (Y_1 \wedge Y_2)$	By (5), (6), M.P.
(8) $\vdash \sim X \supset Y$	By (7), (4), T_6. (Syl.)
(9) $\vdash \sim X \supset X$	By (8), (1), T_6. (Syl.)
(10) $\vdash X$	By (9), T_{24} (b).

T_{42}. (a)
We start the induction with $n = 2$. Well, $\vdash (X_1 \wedge X_2) \supset X_1$ and $\vdash (X_1 \wedge X_2) \supset X_2$, by S_1 and S_2.
Now suppose that $n \geq 2$ and that n is such that for all $i \leq n$, $\vdash (X_1 \wedge X_2 \wedge \cdots \wedge X_n) \supset X_i$. We are to show that for all $i \leq n+1$, $\vdash (X_1 \wedge X_2 \wedge \cdots \wedge X_{n+1}) \supset X_i$. Since $X_1 \wedge X_2 \wedge \cdots \wedge X_{n+1} = (X_1 \wedge X_2 \wedge \cdots \wedge X_{n+1}) \wedge X_{n+1}$, then by S_1 and S_2,

(1) $\vdash(X_1 \wedge X_2 \wedge \cdots \wedge X_{n+1}) \supset X_{n+1}$.

(2) $\vdash(X_1 \wedge X_2 \wedge \cdots \wedge X_{n+1}) \supset (X_1 \wedge X_2 \wedge \ldots . \wedge X_n)$.

Now suppose $i \leq n+1$. Then either $i = n+1$, or $i \leq n$. If the former, then $\vdash(X_1 \wedge X_2 \wedge \cdots \wedge X_{n+1}) \supset X_i$ by (1). Now suppose the latter, i.e. $i \leq n$. Then

(3) $\vdash(X_1 \wedge X_2 \wedge \cdots \wedge X_n) \supset X_i$. (Inductive Hypothesis.)

Hence $\vdash(X_1 \wedge X_2 \wedge \cdots \wedge X_{n+1}) \supset X_i$ [by (2), (3), T_6. (Syl.)].

This completes the induction.

(b)

The proof is similar, using S_5 and S_6 in place of S_1 and S_2.

4. A_1: To show: $X \supset \alpha,\ (X \wedge \alpha_1) \supset Y \vdash X \supset Y$.

(1) $\vdash X \supset \alpha$	Hypothesis.
(2) $\vdash(X \wedge \alpha_1) \supset Y$	Hypothesis.
(3) $\vdash X \supset (\alpha_1 \supset Y)$	By (2), F_3. (Exp.)
(4) $\vdash \alpha \supset \alpha_1$	Fact A_1.
(5) $\vdash X \supset \alpha_1$	By (1), (4), T_6. (Syl.)
(6) $\vdash X \supset Y$	By (5), (3), T_3.

A_2: The proof is similar, using α_2 in place of α_1, and Fact A_2 instead of Fact A_1.

B: To show: $X \supset \beta,\ (X \wedge \beta_1) \supset Y,\ (X \wedge \beta_2) \supset Y \vdash X \supset Y$

(1) $\vdash X \supset \beta$	Hypothesis.
(2) $\vdash(X \wedge \beta_1) \supset Y$	Hypothesis.
(3) $\vdash(X \wedge \beta_2) \supset Y$	Hypothesis.
(4) $\vdash X \supset (\beta_1 \supset Y)$	By (2), F_3. (Exp.)
(5) $\vdash X \supset (\beta_2 \supset Y)$	By (3), F_3. (Exp.)
(6) $\vdash \beta_1 \supset (X \supset Y)$	By (4), T_{13}.
(7) $\vdash \beta_2 \supset (X \supset Y)$	By (5), T_{13}.
(8) $\vdash(\beta_1 \vee \beta_2) \supset (X \supset Y)$	By (6), (7), T_{26}.
(9) $\vdash \beta \supset (\beta_1 \vee \beta_2)$	Fact B.
(10) $\vdash \beta \supset (X \supset Y)$	By (8), (9), T_6. (Syl.)
(11) $\vdash X \supset (\beta \supset Y)$	By (10), T_{13}.
(12) $\vdash X \supset Y$	By (1), (11), T_3.

C: To show: $X \supset Z,\ X \supset {\sim}Z \vdash X \supset Y$

(1) $\vdash X \supset Z$	Hypothesis.
(2) $\vdash X \supset {\sim}Z$	Hypothesis.

(3) $\vdash \sim Z \supset (Z \supset Y)$ By T_{19}.

(4) $\vdash X \supset (Z \supset Y)$ By (2), (3), T_6. (Syl.)

(5) $\vdash X \supset Y$ By (1), (4), T_3.

D: To show: $X \supset \sim X \vdash \sim X$
This is T_{24} (a).

5. In what follows, $\vdash X$ shall mean that X is provable in the system $\mathcal{S}_0$.
   (a) For the axioms of U_1, we are to show that if θ contains some term Y and if $\sim Y$ is also a term, then $\sim C(\theta)$ is provable in $\mathcal{S}_0$.

(1) Y is a term of θ. Hypothesis.

(2) $\sim Y$ is a term of θ. Hypothesis.

(3) $\vdash C(\theta) \supset Y$ By (1), T_{42} (a).

(4) $\vdash C(\theta) \supset \sim Y$ By (2), T_{42} (a).

(5) $\vdash (C(\theta) \supset Y) \wedge (C(\theta) \supset \sim Y)$ By (3), (4), T_8.

(6) $\vdash \sim C(\theta)$ By (5), S_8.

   (b) Rule A_1: We are to show that if α is a term of θ and if $\sim(C(\theta) \wedge \alpha_1)$ [which is $\sim C(\theta, \alpha_1)$] is provable in $\mathcal{S}_0$, so is $C(\theta)$.

(1) α is a term of θ. Hypothesis.

(2) $\vdash \sim(C(\theta) \wedge \alpha_1)$ Hypothesis.

(3) $\vdash C(\theta) \supset \alpha$ By (1), T_{42} (a).

(4) $\vdash \alpha \supset \alpha_1$ By Fact A_1.

(5) $\vdash C(\theta) \supset \alpha_1$ By (3), (4), T_6. (Syl.)

(6) $\vdash \sim C(\theta)$ By (5), (2), T_{38}.

Rule A_2: The proof is similar, using Fact A_2 in place of Fact A_1.

Rule B: We are to show that if β is a term of θ and if both $\sim(C(\theta) \wedge \beta_1)$ and $\sim(C(\theta) \wedge \beta_2)$ are provable in $\mathcal{S}_0$, then $\sim C(\theta)$ is provable in $\mathcal{S}_0$.

(1) β is a term of θ. Hypothesis.

(2) $\vdash \sim(C(\theta) \wedge \beta_1)$ Hypothesis.

(3) $\vdash \sim(C(\theta) \wedge \beta_2)$ Hypothesis.

(4) $\vdash C(\theta) \supset \beta$ By (1), T_{42} (a).

(5) $\vdash \beta \supset (\beta_1 \vee \beta_2)$ Fact B.

(6) $\vdash C(\theta) \supset (\beta_1 \vee \beta_2)$ By (4), (5), T_6. (Syl.)

(7) $\vdash \sim C(\theta)$ By (6), (2), (3), T_{39}.

6. We are given that $\mathcal{T}'$ is an immediate extension of $\mathcal{T}$ and that $\mathcal{T}'$ is bad. We are to show that $\mathcal{T}$ is bad.

Let θ be the branch of $\mathcal{T}$ on which Rule A or Rule B was used to obtain $\mathcal{T}'$.

Case 1. Suppose it was Rule A that was used. Then some α is on θ and $\mathcal{T}'$ resulted from $\mathcal{T}$ by extending branch θ to a branch θ', which is either θ, α_1 or θ, α_2. Since θ' is bad (all branches of $\mathcal{T}'$ are), then so is θ (by Rule A_1 or A_2 of U_1). All other branches of $\mathcal{T}$ are branches of $\mathcal{T}'$, hence are bad. Thus all branches of $\mathcal{T}$ are bad, and so $\mathcal{T}$ is bad.

Case 2. Suppose Rule B was used on θ. Then some β is on θ and $\mathcal{T}'$ is like $\mathcal{T}$, except that in place of the branch θ, the tableau $\mathcal{T}'$ has the two branches θ, β_1 and θ, β_2. Since they are both bad, so is θ (by Rule B of U_1). Again, all other branches of $\mathcal{T}$ are branches of $\mathcal{T}'$, hence are bad. Thus $\mathcal{T}$ is bad.

7. Suppose $\mathcal{T}$ is a closed tableau for X. We are to show that X is bad.

   Let n be the number of applications of Rule A or Rule B used in the construction of $\mathcal{T}$. For any $i \leq n$, let $\mathcal{T}_i$ be the tableau obtained after i applications of rules. Thus $\mathcal{T}_n$ is $\mathcal{T}$. Also, $\mathcal{T}_0$ is simply the formula X. For each $i < n$, the tableau $\mathcal{T}_{i+1}$ is an immediate extension of $\mathcal{T}_i$, so that if $\mathcal{T}_{i+1}$ is bad, so is $\mathcal{T}_i$ (by Problem 6). Also, $\mathcal{T}_n$, being closed, is bad. Hence $\mathcal{T}_{n-1}$ is bad, $\mathcal{T}_{n-2}$ is bad, $\ldots$, $\mathcal{T}_1$ is bad, $\mathcal{T}_0$ is bad. [More formally, for each $i \leq n$, define $P(i)$ to mean that $\mathcal{T}_{n-i}$ is bad. Then $P(0)$ holds (why?), and for each $i < n$, if $P(i)$ holds, so does $P(i+1)$ (why?). Hence by the principle of limited mathematical induction, P holds for all $i \leq n$, which means that $\mathcal{T}_n, \mathcal{T}_{n-1}, \ldots, \mathcal{T}_1, \mathcal{T}_0$, are all bad.]
8. Suppose X is a tautology. Then by the completeness theorem for propositional tableaux, there is a closed tableau for $\sim X$. Hence $\sim X$ is bad (by Problem 7). Thus $\sim\sim X$ is provable in U_1. Hence X is provable in U_1 by Rule N.
9. *Axioms.* First we show that all axioms of U_2 are provable in $\mathcal{S}_0$. Since θ is closed, there is a formula Y such that Y and $\sim Y$ are terms of θ.

(1) Y is a term of θ.	Hypothesis.
(2) $\sim Y$ is a term of θ.	Hypothesis.
(3) $\vdash Y \supset D(\theta)$	By (1), T_{42} (b).
(4) $\vdash \sim Y \supset D(\theta)$	By (2), T_{42} (b).
(5) $\vdash D(\theta)$	By (3), (4) T_{35}.

Re Rule A^0. We are to show that if α is a term of θ, and $D(\theta) \vee \alpha_1$ and $D(\theta) \vee \alpha_2$ are both provable in $\mathcal{S}_0$, then $D(\theta)$ is provable in $\mathcal{S}_0$.

(1) α is a term of θ.	Hypothesis.
(2) $\vdash D(\theta) \vee \alpha_1$	Hypothesis.
(3) $\vdash D(\theta) \vee \alpha_2$	Hypothesis.
(4) $\vdash \alpha \supset D(\theta)$	By (1), T_{42} (b).
(5) $\vdash (\alpha_1 \wedge \alpha_2) \supset \alpha$	Fact A.
(6) $\vdash D(\theta)$	By (4), (2), (3), (5), T_{41}.

Re Rule B_1^0.

(1) β is a term of θ.	Hypothesis.
(2) $\vdash D(\theta) \vee \beta_1$	Hypothesis.
(3) $\vdash \beta \supset D(\theta)$	By (1), T_{42} (b).
(4) $\vdash \beta_1 \supset \beta$	Fact B_1.
(5) $\vdash D(\theta)$	By (3), (2), (4), T_{40}.

Re Rule B_2^0. The proof is the same, using Fact B_2 instead of Fact B_1.

10. Solution to the three parts:

(a) The proof is very similar to the proof in the solution of Problem 8.

Suppose dual tableau $\mathcal{T}'$ is an immediate extension of dual tableau $\mathcal{T}$. Suppose $\mathcal{T}'$ is nice. We are to show that $\mathcal{T}$ is nice. Let θ be the branch of $\mathcal{T}$ on which Rule A^0 or Rule B^0 was used to obtain $\mathcal{T}'$.

Case 1. Suppose dual tableau Rule B^0 was used. Then some β is on θ and θ was extended to a branch θ' of $\mathcal{T}'$ where θ' is either θ, β_1 or θ, β_2. Since θ' is nice (all branches of $\mathcal{T}'$ are nice), then θ is nice (by Rule B_1^0 or B_2^0 of system U_2). All other branches of $\mathcal{T}$ are branches of $\mathcal{T}'$, hence nice. Thus $\mathcal{T}$ is nice.

Case 2. Suppose dual tableau Rule A^0 was used. Then some α is on θ and θ was replaced by the two branches θ, α_1 and θ, α_2. Both of these branches are on $\mathcal{T}'$, hence both are nice. Therefore θ is nice (by Rule A^0 of system U_2). This completes the proof of part (a).

(b) The proof is identical to that of Problem 6, replacing "tableau" by "dual tableau" and "bad" by "nice".

(c) Suppose X is a tautology. Then by the completeness theorem for dual tableaux, there is a closed dual tableau for X. Then by (b) above, X is nice. Hence X is provable in U_2.

Part III
First-Order Logic

8
Beginning First-Order Logic

The Propositional Logic that we have just studied is but the beginning of the logic needed for mathematics and science. The real essence comes in the field known as *First-Order Logic*, which deals with the connectives of Propositional Logic together with the notions of *all* and *some*. Let us first deal with these concepts at an informal level.

In the language of logic, unlike in ordinary English, the word "some" does not have plural connotation; it means only *at least one*, not *two or more*. It means *one or more*. Thus in logic the sentence "some people are good" means nothing more or less than that there exists at least one good person.

As for the notion of *all*, let us recall that the statement "All A's are B's" is to be regarded as automatically true if there are no A's. [The technical term for what I just called "automatically true" is "vacuously true." So, for example, the statement "all unicorns have five legs" is vacuously true, since there are no unicorns.]

Here are a few problems involving the notions of *all* and *some*. We return to that far away cluster of islands on each of which every inhabitant is of one of two types – type T or type F. and where everything said by an inhabitant of type T is true, and everything said by one of type F is false.

Problem 1. On one of my tours of these islands I stopped at a particular island and asked each person living on the island to tell me something about the types of all the people living there. Each one said the same thing: "All of us here are of the same type." Were they really all of the same type, and if so, can it be determined which type they all were?

Problem 2. On the next island that I visited, each one said: "Some of us are of type T and some of us are of type F." What is the composition of that island?

Problem 3. On the next island, I was interested in knowing the smoking habits of the inhabitants, and whether there was any correlation between smoking and truth-telling. They all said the same thing: "Everyone here of type T smokes." What can be deduced about the type T and F distribution, and about smoking habits? [This problem is a bit harder than the first two, and is particularly instructive.]

Problem 4. On the next island, each inhabitant said "Some of us here are of type F and smoke." What can be deduced?

Problem 5. On the next island, all the inhabitants were of the same type, and each one said, "If I smoke, then everyone here smokes." What can be deduced?

Problem 6. On the next island, again all the inhabitants were of the same type, and each one said, "If any one of us here smokes, then I smoke." What can be deduced?

Problem 7. On the next island, again all were of the same type. Each one said, "Some of us smoke, but I don't." What can be deduced?

Problem 8. Suppose I had told you that on that same island, instead of the single statement, "Some of us smoke, but I don't," each inhabitant made two separate statements: (1) Some of us smoke; (2) I don't smoke." What would you conclude? Is the answer the same as that of the last problem?

Introducing ∀ and ∃

In First-Order Logic, we use the letters x, y, z, with or without subscripts, to stand for arbitrary objects of some domain under consideration. What the domain is depends on the application in question. For example, in algebra, the letters x, y, z, would often stand for arbitrary numbers. In geometry, they would often stand for points on a plane. In sociology, they would probably stand for arbitrary people. First-Order Logic is extremely general and is applicable in a wide variety of domains. It has much use in computer science.

Given a property P and any object x, the proposition that x has the property P is symbolized by Px. Now, suppose we wish to say that *every* object has property P, or "all objects have property P." Well, here is where we introduce the symbol "∀," called the *universal quantifier.* The proposition that all objects x have the property P is neatly symbolized "$\forall x Px$" (this is read "for every x, Px" or "for all x, Px").

What about the proposition that *some* x has property P (*some*, in the sense of at least one), or equivalently "there exists an object x that has property P"? This is symbolized "$\exists x Px$" (read "there exists an x having property P"). The symbol "∃" is called the *existential quantifier.*

Incidentally, it is a curious fact that in ordinary English the word *any* sometimes means *some* and sometimes means *all.* For example, if you ask, "Is anyone here," you are obviously not asking whether everyone is here, but whether someone is here. On the other hand, if one says, "Anyone can do this simple task," what is meant is that everyone can do it, not that someone can do it.

Let us now use the quantifiers ∀ and ∃ together with the logical connectives that we have used most frequently from Propositional Logic, namely $\sim$, $\wedge$, $\vee$, $\supset$, $\equiv$.

Let G be the property of being good. Then Gx abbreviates "x is good." $\forall x Gx$ says that everyone is good. $\exists x Gx$ says that at least one individual x is good, or "some people are good." [Remember, *some* means only *at least one.*] How do we symbolically render the proposition that no one is good? One way is $\sim\exists x Gx$ (there does not exist an x such that x is good). Another way is $\forall x(\sim Gx)$ (for every x, x is not good). Let us now abbreviate "x goes to heaven" by Hx. How do we symbolize "All good people go to heaven"? Well, this can be equivalently stated "For every person x, if x is good, then x goes to heaven," and is accordingly symbolized $\forall x(Gx \supset Hx)$. What about the proposition "Only good people go to heaven"? One way is $\forall x(Hx \supset Gx)$. Another way is $\forall x(\sim Gx \supset \sim Hx)$. Still another way is $\sim\exists x(Hx \wedge \sim Gx)$. What about "Some good people go to heaven"? This is obviously symbolized by $\exists x(Gx \wedge Hx)$.

Now let us consider the old saying, "God helps those who help themselves." It seems that there is some ambiguity there – does it mean that God helps *all* people who help themselves, or does it mean that God helps *only* those who help themselves, or does it mean that God helps all and only those who help themselves? Well, let us abbreviate "God" by "g" and x helps y by "xHy." Thus God helps *all* those who help themselves would be symbolized by $\forall x(xHx \supset gHx)$. The proposition that God helps *only* those who help themselves would be rendered $\forall x(gHx \supset xHx)$. As for the proposition that God helps all and only those who help themselves, a rendition is $\forall x((xHx \supset gHx) \wedge (gHx \supset xHx))$, or more simply, $\forall x(gHx \equiv xHx)$.

Now for some more translations.

Problem 9. Let h stand for Holmes (Sherlock Holmes) and let m stand for Moriarty. Let us abbreviate "x can catch y" by "xCy". Give symbolic renditions of the following:

(a) Holmes can catch anyone who can catch Moriarty.
(b) Holmes can catch anyone whom Moriarty can catch.
(c) Holmes can catch anyone who can be caught by Moriarty.
(d) If anyone can catch Moriarty, then Holmes can.
(e) If everyone can catch Moriarty, then Holmes can.
(f) Anyone who can catch Holmes can catch Moriarty.
(g) No one can catch Holmes unless he can catch Moriarty.
(h) Everyone can catch someone who cannot catch Moriarty.
(i) Anyone who can catch Holmes can catch anyone whom Holmes can catch.

Problem 10. Let us symbolize "x knows y" by "xKy". Give symbolic renditions of the following:

(a) Everyone knows someone.
(b) Someone knows everyone.
(c) Someone is known by everyone.

(d) Every person x knows someone who doesn't know x.
(e) There is someone x who knows everyone who knows x.

Problem 11. Let Dx abbreviate "x can do it" and let "b" abbreviate Bernard. Let "$x = y$" abbreviate "x is identical with y". Give symbolic renditions of the following:

(a) Bernard, if anyone, can do it.
(b) Bernard is the only one who can do it.

Problem 12. Let us now consider some examples from arithmetic. Here "number" shall mean *natural number*, i.e. 0 or some positive whole number. The usual abbreviation of "x is less than y" is "$x < y$", and the abbreviation of "x is greater than y" is "$x > y$". Express the following statements symbolically:

(a) For every number there is a greater number.
(b) Every number other than 0 is greater than some number.
(c) 0 is the one and only number having the property that no number is less than it.
(d) Without using the identity symbol, $=$, but using $<$ and/or $>$, express the properties that x is equal to y. Express the property that x is unequal to y.

Interdependence of ∀ and ∃

Problem 13. Again we consider a property P of objects and symbolize the proposition that x has the property P by Px. The property that all objects x have property P is symbolized by $\forall x Px$. However, it is possible to symbolize that all x have property P without using the universal quantifier $\forall$, but instead using the existential quantifier $\exists$, together with some of the connectives of Propositional Logic. How? Thus $\forall$ is definable from $\exists$ and propositional connectives. How?

Relational Symbols

Consider a relation R between two objects. The proposition that x stands in the relation R to y is symbolized by Rx, y, or sometimes by xRy. Now consider a relation R of three arguments, i.e. a relation between three objects x, y and z (such as $x + y = z$). The proposition that x, y, z stands in the relation R is symbolized Rx, y, z. Similarly with a relation R of n arguments, for $n \geq 3$. Thus, $Rx_1, x_2, \ldots, x_n$ expresses the proposition that $x_1, x_2, \ldots, x_n$ stands in the relation R. Relations of one argument are called *properties*.

Formulas of First-Order Logic

For First-Order Logic (also called *quantification theory*) we shall use the following symbols:

(a) The symbols of Propositional Logic, other than propositional variables.
(b) $\forall$ (read "for all"),
$\exists$ (read "there exists").
(c) A denumerable list of symbols called *individual variables.*
(d) A denumerable list of symbols called *individual parameters.*
(e) For each positive integer n, a set of symbols called *n-ary predicates*, or *predicates of degree n.*

The term *variable* shall henceforth mean *individual variable* (not to be confused with the propositional variables of Propositional Logic). We shall use lower-case letters x, y, z, with or without subscripts, to denote arbitrary variables. We shall use letters a, b, c, with or without subscripts, to denote individual parameters (henceforth called just "parameters"). We shall use upper-case letters P, Q, R, with or without subscripts, to denote predicates, whose degrees will always be clear from the context. We shall use the term "individual symbols" collectively for (individual) variables and parameters.

Atomic Formulas

By an *atomic formula* we shall mean a predicate of degree n followed by n individual symbols.

Formulas

Starting with the atomic formulas, we build the set of all formulas (of First-Order Logic) by the formation rules of Propositional Logic, together with the rule that for any formula F, and any variable x, the expressions $\forall x F$ and $\exists x F$ are formulas. The rules are thus the following.

(1) Every atomic formula is a formula.
(2) For any pair of formulas F and G, the expressions $\sim F$, $(F \wedge G)$, $(F \vee G)$, $(F \supset G)$, and $(F \equiv G)$, are formulas.
(3) For any formula F and variable x, the expressions $\forall x F$ (called the *universal quantification* of F with respect to x) and $\exists x F$ (called the *existential quantification* of F with respect to x) are formulas.

No expression is a formula unless its being so is a consequence of the above three rules.

Degrees of Formulas

By the *degree* of a formula is meant the number of occurrences of the symbols $\sim$, $\wedge$, $\vee$, $\supset$, $\equiv$, $\forall$, and $\exists$ in it.

Free and Bound Occurrences of Variables

We now must deal with what some (including the author) regard as the most annoying aspect of First-Order Logic!

Before defining the important notion of free and bound occurrences of individual variables, let us look at some examples.

In the arithmetic of the natural numbers, consider the following equation:

$$x = 5y.$$

This equation as it now stands is neither true nor false, but becomes true or false when we assign values to the variables x and y. For example, if we take x to be 15 and y to be 3, we have a truth. If we take x to be 12 and y to be 9, we have an obvious falsehood. The important thing now is that the truth or falsity of the above equation depends on both a choice of value for x and a choice of a value for y. This reflects the fact that both x and y occur *freely* in the equation.

Now consider the following:

$$\exists x(y = 5x).$$

The truth or falsity of the above depends on y, but not on any choice for x. Indeed, we could restate the above in a form in which the variable x does not even occur, namely "y is divisible by 5." This reflects the fact that y has a free occurrence in the above, but x does not; we say that x is *bound* in that equation.

Suppose x has an occurrence in a formula F. If one puts $\forall x$ or $\exists x$ if front of F, then all occurrences of x in $\forall x F$ or $\exists x F$ become bound. Thus all occurrences of x in $\forall x F$ and $\exists x F$ are bound occurrences. Here are the precise rules determining freedom and bondage:

(1) In an atomic formula, all occurrences of variables are free.

(2) The free occurrences of a variable x in $\sim F$ are the same as those in F. The free occurrences of x in $(F \wedge G)$ are those of F and those of G. Likewise with $\vee$, $\supset$, and $\equiv$ instead of $\wedge$.

(3) All occurrences of a variable x in $\forall x F$ are bound (not free), but for any variable y distinct from x, the free occurrences of y in $\forall x F$ are those in F itself. Similarly with $\exists$ in place of $\forall$.

It is possible that a variable may have both a free and a bound occurrence in the same formula. For instance, $Px \supset \forall x Qx$ [the occurrence of x following P is free, while the occurrence following Q is bound].

A formula is called *closed* if it has no occurrences of free variables, and *open*, otherwise. Closed formulas are also called *sentences*.

Substitution

For every formula F, variable x, and parameter a, by F^x_a is meant the results of replacing every free occurrence of x in F by a. This substitution operation is in accordance with the following conditions:

(1) If F is atomic, then F^x_a is the result of substituting a for *every* occurrence of x in F.
(2) For any pair of formulas F and G, $(F \wedge G)^x_a = F^x_a \wedge G^x_a$. Similarly with $\vee$, $\supset$, and $\equiv$ instead of $\wedge$. $(\sim F)^x_a = \sim(F^x_a)$.
(3) $[\forall x F]^x_a = \forall x F$ and $[\exists x F]^x_a = \exists x F$, but for any variable y distinct from x, $[\forall x F]^y_a = \forall x[F]^y_a$ and $[\exists x F]^y_a = \exists x[F]^y_a$.

The following notation is convenient. We let $\varphi(x)$ be any formula with x as a free variable. Then for any parameter a by $\varphi(a)$ is meant the result of substituting a for all free occurrences of x in $\varphi(x)$. Thus $\varphi(a)$ is $[\varphi(x)]^x_a$. Even if x doesn't occur free in $\varphi(x)$, we can use the notation $\varphi(x)$ [meaning φ is a formula which may have a free variable x] and we will still have $\varphi(a) = [\varphi(x)]^x_a$, because by definition in this case the substitution of a for every free occurrence of x in $\varphi(x)$ effects nothing, just yields $\varphi(x)$ once more.

Note: Because the phrase "if and only if" occurs so frequently in mathematics the mathematician Paul Halmos suggested using the abbreviation "iff" for it, and we shall follow him very often in the remainder of this text.

Interpretations and Valuations

Pure formulas are those containing no parameters. For the moment, we shall consider pure closed formulas (i.e. formulas with no parameters and with no occurrences of free variables).

U-formulas

In defining the notion of an *interpretation*, the first thing we must do is to select some non-empty set U called the *domain* of the interpretation. Without loss of generality, we can assume that the elements of U are symbols, but distinct from all symbols of First-Order Logic so far considered. The only thing that matters in the chosen domain is the number of its elements. If the elements of U are not symbols, but either extra-linguistic entities such as numbers, we could assign to each element of U a symbol to name it, with the understanding that distinct elements of U have distinct names, and then U will be the same size as its set of names. But to simplify matters, we will assume that the elements of U are themselves symbols. Note that we are not assuming here that there are only finitely or denumerably many symbols when we say this.

We now wish to define the notion of a formula with constants in U – more briefly, a U-formula. By an *atomic* U-formula we shall mean an expression consisting of a predicate of some degree n, followed by n symbols, each of which is either a variable or an element of U. Thus an atomic U-formula is an expression of the form $Pe_1, \ldots, e_n$, where P is a predicate of degree n and each e_i is either a variable or a symbol of U. Thus an atomic U-formula is like an atomic formula except that it has elements of U in place of parameters. If an atomic formula $Pe_1, \ldots, e_n$ is *closed*, then of course each e_i is a symbol of U (since all occurrences of variables in an atomic formula are free, hence if there are no free variables in it, then there are no variables at all in it).

What we have called an *interpretation* in Propositional Logic will now be called a *propositional interpretation*. In First-Order Logic, an interpretation I is specified by first selecting a non-empty domain U and then assigning to each predicate P of degree n, an n-ary relation of elements of U. We let $I(P)$ be the relation assigned to P under I. Under the interpretation I, every closed U-formula (and this includes all closed pure formulas in which no constants of U appear) becomes either true or false, according to the following rules.

(1) For any n-ary predicate P and elements $u_1, \ldots, u_n$ of U, the atomic U-formula $Pu_1, \ldots, u_n$ is true (under I) iff the n elements $u_1, \ldots, u_n$ stand in the relation assigned to P under I.
(2) As in Propositional Logic, $\sim F$ is true (under I) iff F is not true. $(F \wedge G)$ is true iff F and G are both true. $(F \vee G)$ is true iff at least one of F and G are true. $(F \supset G)$ is true iff either F is not true or G is true. $(F \equiv G)$ is true iff both F and G are true or both F and G are false.
(3) $\forall x\varphi(x)$ is true iff $\varphi(u)$ is true for every element u of U. $\exists x\varphi(x)$ is true iff $\varphi(u)$ is true for at least one element u of U.

This concludes the definition of truth under *an interpretation of predicates for closed pure formulas*.

For a formula F with free variables or/and parameters, by *an interpretation I in a domain U of F* is meant an interpretation of the predicates of F, together with an assignment of an element of U to each free variable and each parameter of F. We then take F to be true under I if F' is true under I, where F' is the result of replacing each free variable and parameter of F by the element of U assigned to it under I.

This concludes the definition of truth of formulas with or without free variables or parameters, under an interpretation in a domain U.

In First-Order Logic, a formula is called *valid* in the domain U if it is true under *every* interpretation in the domain U, and is called *satisfiable* in U iff it is true under at least one interpretation in U. A formula is called *valid* if it is valid in every non-empty domain, and *satisfiable* if it is satisfiable in at least one non-empty domain. A set of formulas is called simultaneously satisfiable if there is at least one domain such that all the formulas of the set are true in at least one interpretation in that domain.

If a pure closed formula X is satisfiable in a domain U_1, then it is satisfiable in any larger domain U_2, that is, in any domain U_2 such that U_1 is a proper subset of U_2. This can be seen as follows.

Let I_1 be an interpretation of X in the domain U_1 under which X is true. We wish to construct an interpretation I_2 of U_2 in which X is true.

Let r be any element of U_1. For any element e of U_2 define e' as follows: If e is in U_1, take e' to be e. If e is in U_2 but not in U_1, take e' to be the element r of U_1. For any U_2-formula F, let F' be the result of replacing each element e of U_2 by e'. Now we are ready to define the interpretation I_2 in U_2 : For each predicate P of degree n in X, take $I_2(P)$ to be the set of all n-tuples $(e_1, \ldots, e_n)$ of elements of U_2 such that $Pe'_1, \ldots, e'_n$ is true under I_1. By complete mathematical induction on degrees, it can be shown that for every U_2-formula F, the formula F is true under I_2 iff F' is true under I_1 (exercise below). In particular, X is true under I_2 iff X is true under I_1, but since X is true under I_1, then X is true under I_2.

Exercise. Carry out the induction.

Problem 14. Suppose domain U_1 is a proper subset of U_2. It is not conversely true that if a closed pure formula is satisfiable in U_2 then it is necessarily satisfiable in U_1. As a counterexample, exhibit a closed pure formula that is satisfiable in any domain with two elements, but is not satisfiable in any domain with only one element.

Problem 15. Exhibit a pure closed formula that is satisfiable in a denumerable domain, but not satisfiable in any finite domain.

Problem 16. Can you find a formula that is satisfiable in a non-denumerable domain, but not satisfiable in any denumerable domain?

Tautologies

A formula X of First-Order Logic is said to be an *instance* of a formula Y of Propositional Logic if X is obtainable from Y by substituting formulas of First-Order Logic for the propositional variables of Y. For instance $\forall x Qx \vee \sim\exists y Py$ is an instance of the propositional formula $p \vee \sim q$ (it is obtainable by substituting $\forall x Qx$ for p and $\sim\exists y Py$ for q). Now, a first-order formula X is called a *tautology* if it is an instance of a tautology of Propositional Logic. For example, $\forall x\, Qx \vee \sim\forall x\, Qx$ is a tautology, because it is an instance of the propositional tautology $p \vee \sim p$. Even if one did not know what the symbol $\forall$ meant, but knew the meanings of $\vee$ and $\sim$, one would know that $\forall x\, Qx \vee \sim\forall x\, Qx$ must be true, because for *any* proposition, either it or its negation must be true. Another tautology is $(\forall x\, Px \wedge \forall x\, Qx) \supset \forall x\, Px$, because it is an instance of $(p \wedge q) \supset p$. However, the formula $(\forall x\, Px \wedge \forall x\, Qx) \supset \forall x(Px \wedge Qx)$, though valid, is *not* a tautology! It is valid because if *every* element has property P and every element has property Q, then every element has properties P and Q. But the formula is not an instance of any tautology of Propositional Logic. To realize the validity of the formula, one must know what the symbol $\forall$ means. For example, if one reinterpreted $\forall$ to mean "there exists" instead of "for all," the formula wouldn't always be true (if some element has property P and some element has property Q, it doesn't follow that some element has both properties P and Q). All tautologies are of course valid, but they constitute only a fragment of the valid formulas of First-Order Logic.

An Axiom System for First-Order Logic

There are several axiom systems for First-Order Logic in the literature. Some of them extend complete axiom systems for Propositional Logic to axiom systems for First-Order Logic by adding axioms and inference rules for the quantifiers. Other axiom systems, instead of taking axioms for the propositional part, simply take all tautologies as axioms. This is perfectly legitimate, since one can effectively tell whether a formula is a tautology by a truth table. This is the course we will take for now. [In a later chapter it will prove more convenient to have axioms for the propositional part, such as those of Chapter 7.]

Here is our axiom system for First-Order Logic, which we will name S_1 (the subscript 1 reminding us that this is a proof system for *First*-Order Logic):

Axioms

Group 1. All tautologies.
Group 2. (a) All sentences $\forall x\varphi(x) \supset \varphi(a)$.
(b) All sentences $\varphi(a) \supset \exists x\varphi(x)$.

Inference Rules.

I. Modus Ponens $\dfrac{X, \quad X \supset Y}{Y}$.

II. (a) $\dfrac{\varphi(a) \supset X}{\exists x\varphi(x) \supset X}$, (b) $\dfrac{X \supset \varphi(a)}{X \supset \forall x\varphi(x)}$,

where X is closed and a is a parameter that does not occur in either X or $\varphi(x)$.

Problem 17. Prove that the axiom system $\mathcal{S}_1$ for First-Order Logic is correct.

Thus the system is correct. The amazing thing is that the system is *complete*, i.e. that all valid formulas are provable in it! This is a major result of First-Order Logic, and is really due to Kurt Gödel, and is knows as *Gödel's Completeness Theorem.* Gödel proved this theorem for a system closely related to the above system in 1930 as his Ph.D, thesis at the University of Vienna. Only a year later he was to prove his even more famous incompleteness theorem (which applies to any formal axiomatic system that is powerful enough to describe the arithmetic of the natural numbers). This incompleteness theorem is a subject that we will look at in Part IV of this book.

The proof we will give of completeness is much simpler than the original proof of Gödel. It is simplified by the use of first-order tableaux, which is the subject of the next chapter.

Solutions to the Problems of Chapter 8

1. Since they all said the same thing about the nature of the island, then of course they are all of the same type. Since they all truthfully said that they were all of the same type, then all of them are of type T.
2. Again, they all said the same thing, hence they are all of the same type. Thus what each one said was false, and so they are all of type F.
3. As in the previous two problems, they are all of the same type, since they all said the same thing about the island. Suppose they were all of type F. Then their statements were false since they always lie, and thus it is false that all those of type T smoke. But the only way that could be false is if there were at least one inhabitant of type T who doesn't smoke, which contradicts the assumption that all are of type F. Thus the assumption cannot be true, and since all must be of the same type, they must all be of type T. It then further follows that since their statements were true, all of them smoke. Thus the solution is that all of them are of type T and all of them smoke.
4. Again, they are all of the same type. If they were of type T, they wouldn't have said that some are of type F and smoke, since this would imply that some are of type F. Thus they cannot be of type T; they are all of type F. It further follows that their statements are false, which means it is not the case that some are of type F and smoke. Thus they are all of type F and none of them smoke.
5. We are given that all are of the same type. Suppose they are all of type F. Then each one's statement is false, which means that the islander saying it smokes but not all of them smoke. It is clearly impossible that each islander smokes, yet not all of them smoke. Hence the assumption that they are all of type F leads to a contradiction. Thus they are all of type T. Hence each one's statement is true, which means that for each one, either that one doesn't smoke, or all of them

smoke. Consequently none of them smokes or all of them smoke, and there is no way to tell which. Thus all that can be deduced is that all are of type T and either all smoke or none of them smoke.

6. Again, we're given that all the inhabitants are of the same type. Each inhabitant claims that if any inhabitant smokes, then he does. Here the world "any" means *some*. Thus each is claiming that if at least one inhabitant smokes, then he does. If the claim were false, that would mean that some inhabitant smokes, but the speaker does not, which means (since every inhabitant says this) that some inhabitant smokes, but each speaker does not, which is an obvious contradiction. Thus all the statements were true, so as in the last problem, all are of type T. Hence also, as in the last problem, either none of them smoke or all of them smoke, and there is no way to tell which.
7. Again, we are given that all are of the same type. They couldn't all be of type T, because if their statements were true, then some of them smoke, yet each one doesn't, which is absurd. Therefore they are all of type F. Since their statements are false, it follows that for each inhabitant x, either it is false that some inhabitants smoke or it is false that x doesn't smoke; in other words either one of them smokes or x smokes. It could be that none of them smoke. If that alternative doesn't hold, then each x smokes, which means that all of them smoke. Thus all of them are of type F, and either all of them smoke or none of them smoke, and there is no way to tell.
8. I didn't tell you that, but *if* I had, what you should conclude is that I must have either been lying or was mistaken, since it is impossible that all the inhabitants could make both of these statements separately. Here is why:

   Suppose each person x said:

   (1) Some of us smoke.
   (2) I don't smoke.

   We are again given that all are of the same type. Suppose they were of type T. Then their statements were both true. Then by (1), some of them smoke. And by (2), each x doesn't smoke. This is clearly a contradiction.

   Suppose they were of type F. Then their statements were both false. Since (1) is false, then none of them smoke. Yet each x falsely said that he, x, doesn't smoke, which means that every x smokes. This is also a contradiction. Thus it cannot be that all the inhabitants made the reported statements (1) and (2).

   Note. This problem along with the one before it provide an interesting example of how an islander of type F can assert the conjunction of two statements, yet cannot assert each statement separately. And this is why the solutions to these last two problems are different.
9. (a) $\forall x(xCm \supset hCx)$
   (b) $\forall x(mCx \supset hCx)$

(c) Same as (b)
(d) $\exists x(xCm) \supset hCm$
(e) $\forall x(xCm) \supset hCm$
(f) $\forall x(xCh \supset xCm)$
(g) Same as (f)
(h) $\forall x \exists y(xCy \wedge \sim yCm)$
(i) $\forall x(xCh \supset \forall y(hCy \supset xCy))$; alternatively, $\forall x \forall y((xCh \wedge hCy) \supset xCy)$

10. (a) $\forall x \exists y(xKy)$
(b) $\exists x \forall y(xKy)$
(c) $\exists x \forall y(yKx)$
(d) $\forall x \exists y(xKy \wedge \sim yKx)$
(e) $\exists x \forall y(yKx \supset xKy)$

11. (a) $\exists x Dx \supset Db$
Alternatively, $\forall x(Dx \supset Db)$
(b) $Db \wedge \forall x(Dx \supset (x = b))$
Alternatively, $\forall x(Dx \equiv (x = b))$

12. (a) $\forall x \exists y(y > x)$
(b) $\forall x(\sim (x = 0) \supset \exists y(x > y))$
(c) $\sim \exists y(y < 0) \wedge \forall x(\sim \exists y(y < x) \supset (x = 0))$
Alternatively, $\forall x(\sim \exists y(y < x) \equiv (x = 0))$
(d) x is equal to y can be expressed by $\sim(x < y) \wedge \sim(y < x)$
x is unequal to y can be expressed by $(x < y) \vee (y < x)$

13. To say that every x had property P is equivalent to saying that there does not exist an x that doesn't have property P. Thus $\forall x Px$ is equivalent to $\sim\exists x \sim Px$. Also, to say that there exists an x that has property P is equivalent to saying that it is not the case that every x doesn't have property P. Thus $\exists x Px$ is equivalent to $\sim\forall x \sim Px$.

14. Let F be the conjunction of the following two formulas:

F_1: $\forall x \exists y Rxy$
F_2: $\sim\exists x Rxx$

Thus $F = \forall x \exists y Rxy \wedge \sim\exists x Rxx$. Now take a domain $\{a, b\}$ of two *distinct* objects a and b. Interpret Rxy to mean that $x \neq y$ (x is not equal to y). Under this interpretation, F_1 holds, since if $x = a$, then there is some y, namely b, such that $a \neq y$; and if $x = b$, then there is some y, namely a, such that $b \neq y$. Thus for each x in the domain, there is some y such that Rxy. Thus $\forall x \exists y Rxy$ is true, and thus F_1 holds. Also, F_2 holds, since $\sim a \neq a$ and $\sim b \neq b$, which is what $\sim Raa$ and $\sim Rbb$ means under the interpretation. And so for all x in the domain, $\sim Rxx$. Consequently $\sim\exists x Rxx$ holds. Thus both F_1 and F_2 hold, so that their conjunction F also holds in the same domain. Thus F is satisfiable in the two-element domain $\{a, b\}$.

Now, to show that F is not satisfiable in a one-element domain, let D be any domain having at least one element e. Suppose F is satisfiable over D. Regardless of how we interpret R, both F_1 and F_2 are true under the interpretation. Since F_1 is true, then Rey is true for at least one element y of D. This y cannot be e itself, by F_2! Thus y is distinct from e, and so D contains at least two distinct elements e and y. Thus F is not satisfiable in a domain with only one element.

15. Let F_1, F_2 and F_3 be the following formulas (F_1 and F_2 are the same as in the last problem):

    F_1: $\forall x \exists y Rxy$
    F_2: $\sim\exists x Rxx$
    F_3: $\forall x \forall y \forall z((Rxy \wedge Ryz) \supset Rxz)$

    Let F be the conjunction $F_1 \wedge F_2 \wedge F_3$. Well, F is satisfiable in the denumerable domain of all the natural numbers, taking Rxy to mean $x < y$ (x is less than y).

    (1) Of course, for any number x there is a number y such that x is less than y. Thus F_1 holds.
    (2) No number is less than itself; hence F_2 holds.
    (3) If x is less than y, and y is less than z, then of course x is less than z, and so F_3 holds.

    Next we show that if D is any non-empty domain in which F can be satisfied, then D must contain infinitely many elements. Well, let R be any relation of elements of D for which F is true. Since D is non-empty, it contains at least one element e_1. By F_1 the element e_1 stands in the relation R to some element e_2. This element e_2 must be distinct from e_1 by F_2. Next Re_2e_3 holds for some element e_3 (by F_1) and e_3 must be distinct from e_2 (again by F_2). Also, since Re_1e_2 and Re_2e_3 hold, then Re_1e_3 holds (by F_3), so that e_3 is distinct from e_1. Now we have three distinct elements e_1, e_2, and e_3. Now we have Re_3e_4, for some element e_4, which by similar arguments must be different from e_1, e_2, and e_3. Then Re_4e_5 holds from some new element e_5, and so forth. Thus we get an infinite sequence $e_1, e_2, \dots, e_n \dots$. of distinct elements.

16. The answer to the question is NO! You cannot find such a formula, because there is none! This important fact is the famous theorem of Leopold Löwenheim, which is that if a formula is satisfiable at all, then it is satisfiable in a denumerable domain. This was later improved by Thoralf Skolem in what is now known as the *Löwenheim-Skolem Theorem*, which is that for any

denumerable set S of formulas, if S is simultaneously satisfiable, then it is simultaneously satisfiable in a denumerable domain. Thus no formula, not even a denumerable set of formulas, can force the domain of an interpretation to be non-denumerable. This is one of the major results of First-Order Logic, and will be proved in the next chapter.

17. The following is a proof the Axiom System for First-Order Logic presented in this chapter is *correct.*

Axioms

Group 1. All tautologies.
Group 2. (a) All sentences $\forall x\varphi(x) \supset \varphi(a)$.
(b) All sentences $\varphi(a) \supset \exists x\varphi(x)$.

(1) Obviously all tautologies are valid.
(2a) To say that $\forall x\varphi(x) \supset \varphi(a)$ is valid is to say that for every interpretation I in a non-empty domain U, the sentence $\forall x\varphi(x) \supset \varphi(e)$ is true for every element e of U assigned to the parameter a. This is obviously so.
(2b) Consider any interpretation I of the predicates and parameters of $\varphi(x)$ (which do not include the parameter a) in some domain U. We are to show that whatever element e of U we assign to the parameter a, the sentence $\varphi(e) \supset \exists x\varphi(x)$ will be true (under I). Well, if $\varphi(e)$ is false, then $\varphi(e) \supset \exists x\varphi(x)$ is vacuously true. On the other hand, if $\varphi(e)$ is true, the so is $\exists x\varphi(x)$, and hence so is $\varphi(e) \supset \exists x\varphi(x)$.

Now for the *Inference Rules.*

I. Modus Ponens $\dfrac{X,\ X \supset Y}{Y}$.

II. (a) $\dfrac{\varphi(a) \supset X}{\exists x\varphi(x) \supset X}$, (b) $\dfrac{X \supset \varphi(a)}{X \supset \forall x\varphi(x)}$,

where X is closed and a is a parameter that does not occur in either X or $\varphi(x)$.

Of course Rule I (Modus Ponens) is correct (i.e. preserves validity).
Now for Rule II.
(a) Suppose $\varphi(a) \supset X$ is valid and a does not occur in either X or $\varphi(x)$. Then under any interpretation I in a domain U, and any element e of U, the sentence $\varphi(e) \supset X$ is true (under I). We are to show that $\exists x\varphi(x) \supset X$ is true (under I). Well, suppose $\exists x\varphi(x)$ is true. Then for some element e of U, the sentence $\varphi(e)$ is true, and since $\varphi(e) \supset X$ is true, then X must be true, and thus $\exists x\varphi(x) \supset X$ is true (under I). Of course, if $\exists x\varphi(x)$ is false, then $\exists x\varphi(x) \supset X$ is true (under I) in this case as well.

(b) Suppose $X \supset \varphi(a)$ is valid. We are to show that $X \supset \forall x\varphi(x)$ is valid. Consider any interpretation I in a domain U. By what it means for $X \supset \varphi(a)$ to be valid, for every element e of U, the sentence $X \supset \varphi(e)$ is true under I. To show that $X \supset \forall x\varphi(x)$ is true under I, suppose X is true. Since $X \supset \varphi(e)$ is true for every e in U, and X is true, $\varphi(e)$ is true for every element e of U. Thus $\forall x\varphi(x)$ is true under I. Thus $X \supset \forall x\varphi(x)$ is true under I. Of course, if X is false under I, then $X \supset \forall x\varphi(x)$ must also be true under I, since a false statement always implies any statement.

9

First-Order Logic: Main Topics

First-Order Tableaux

The tableaux which will be defined here have far-reaching consequences, and may well be of even greater significance than axiom systems for First-Order Logic. Indeed, the completeness of the axiom systems is neatly derivable from the completeness theorem for first-order tableaux.

First-order tableaux use the eight tableau rules of Propositional Logic together with the four rules for the quantifiers that will be given shortly. But first for some examples.

Suppose we wish to prove the formula $\exists x Px \supset \sim\forall x \sim(Px \vee Qx)$. As with Propositional Logic, we begin the tableaus with F followed by the formula we wish to prove:

$$(1)\quad F\ \exists x Px \supset \sim\forall x \sim(Px \vee Qx)$$

Next we use rules from Propositional Logic to extend the tableau as follows. [As in some of our Propositional Logic proofs, to help the reader here we put to the right of each line the number of the line from which it was inferred.]

(2) $T\ \exists x Px$	By (1).
(3) $F \sim\forall x \sim(Px \vee Qx)$	By (1).
(4) $T\ \forall x \sim(Px \vee Qx)$	By (3).

There is no rule from Propositional Logic that is now applicable, so we turn to working on the quantifiers. By line (2) there is at least one x such that Px; we let the parameter a be such an x, and we add the following line to the proof:

(5) $T\ Pa$	By (2).

Next we look at line (4), which says that whatever x we take, it is not the case that Px or Qx. In particular, it is not the case that Pa or Qa, and so we add:

(6) $T \sim(Pa \vee Qa)$	By (4).

At this point lines (5) and (6) constitute a clean inconsistency in Propositional Logic, and we could stop right here, or alternatively, close the tableau using only rules of Propositional Logic thus:

(7) $F\ Pa \vee Qa$ By (6).

(8) $F\ Pa$ By (7).

[Line (8) clashes with line (5).]

In the future, to save unnecessary labor, whenever we obtain a branch that contains an inconsistency in Propositional Logic, we can put a bar under it and treat it as closed, since we know that it can be closed using tableau rules for Propositional Logic. Let us consider another example. Let us prove the formula $(\forall x Px \wedge \exists x(Px \supset Qx)) \supset \exists x Qx$.

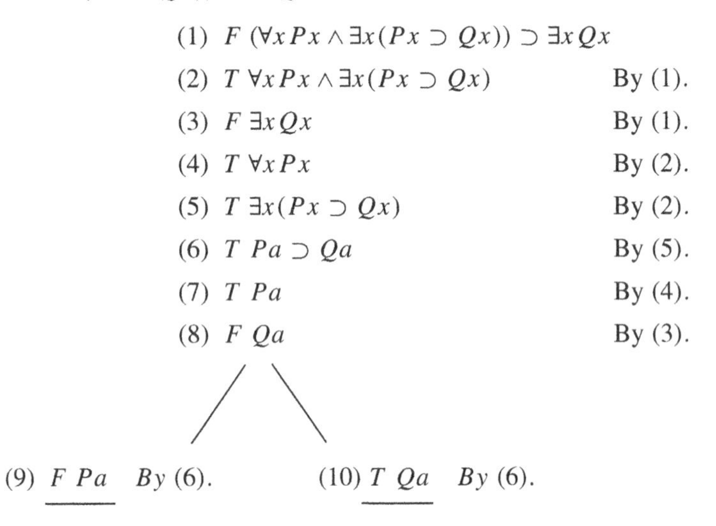

(1) $F\ (\forall x Px \wedge \exists x(Px \supset Qx)) \supset \exists x Qx$

(2) $T\ \forall x Px \wedge \exists x(Px \supset Qx)$ By (1).

(3) $F\ \exists x Qx$ By (1).

(4) $T\ \forall x Px$ By (2).

(5) $T\ \exists x(Px \supset Qx)$ By (2).

(6) $T\ Pa \supset Qa$ By (5).

(7) $T\ Pa$ By (4).

(8) $F\ Qa$ By (3).

(9) $F\ Pa$ By (6). (10) $T\ Qa$ By (6).

Explanation. Lines (2), (3), (4), and (5) were obtained by tableau rules of Propositional Logic. Line (5) says that $Px \supset Qx$ holds for at least one x, and so we let a be such an x, and thus get line (6). Line (4) says that Px holds for every x, and so, in particular, Pa holds, which gives us line (7). Line (3) says that it is false that there is any x such that Qx, hence, in particular Qa is false, which gives line (8). Then line (6) branches to the formulas of lines (9) and (10), and the tableau then closes because there are contradictions on both of its branches.

Tableau Rules for the Quantifiers

We have four quantifier rules, one for each of $T\ \forall x\varphi(x)$, $T\ \exists x\varphi(x)$, $F\ \forall x\varphi(x)$, $F\ \exists x\varphi(x)$. None of these rules involve branching.

Rule $T\ \forall$. From $T\ \forall x\varphi(x)$ we may directly infer $T\ \varphi(a)$, where a is any parameter.

Rule T ∃. From $T\ \exists x\varphi(x)$ we may infer $T\ \varphi(a)$, *provided that a does not yet occur on the tableau.*

Here is the reason for the proviso in italics just above: Suppose that in the course of a proof we show that some x has a certain property P. We can then say, "Let a be such an x." Now suppose we later show that there is an x having some other property Q. We cannot legitimately say, "Let a be such an x," because we have already committed the symbol "a" to being the name of some x having property P, and we do not know that there is *any* x having *both* property P and property Q! Thus we must take a new symbol b and say, "Let b be an x having property Q."

Rule F ∀. [This is similar to Rule T ∃.] From $F\ \forall x\varphi(x)$ we may directly infer $F\ \varphi(a)$, provided that a is new to the tableau.

Here $F\ \forall x\varphi(x)$ says that it is false that $\varphi(x)$ holds for every x, which is equivalent to saying that there is at least one x such that $\varphi(x)$ is false. We let a be such an x and write $F\ \varphi(a)$, but again, a must be new to the tableau for the same reason as for Rule T ∃.

Rule F ∃. From $F\ \exists x\varphi(x)$ we may infer $F\ \varphi(a)$ for any parameter a (no restriction necessary).

This time $F\ \exists x\varphi(x)$ says that it is false that there is any x such that $\varphi(x)$ is true, or, in other words, that $\varphi(x)$ is false *for every* x, and thus $F\ \varphi(a)$ holds, whatever a may be.

Let us now review the four rules in schematic form:

Rule T ∀ $\quad \dfrac{T\ \forall x\varphi(x)}{T\ \varphi(a)}$ (a is any parameter) $\qquad$ *Rule F ∃* $\quad \dfrac{F\ \exists x\varphi(x)}{F\ \varphi(a)}$ (a is any parameter)

Rule T ∃ $\quad \dfrac{T\ \exists x\varphi(x)}{T\ \varphi(a)}$ (a must be new) $\qquad$ *Rule F ∀* $\quad \dfrac{F\ \forall x\varphi(x)}{F\ \varphi(a)}$ (a must be new)

Rules T ∀ and F ∃ are collectively called *universal* rules (in spite of the existential symbol ∃, the formula $F\ \exists x\varphi(x)$ asserts the *universal* fact that for *every* element a, it is false that $\varphi(a)$ holds). The rules T ∃ and F ∀ are called *existential* rules (the formula $F\ \forall x\varphi(x)$ asserts the *existential* fact that $\varphi(a)$ is false for at least one element a).

For *unsigned* formulas, the quantification rules are the following:

Rule ∀ $\quad \dfrac{\forall x\varphi(x)}{\varphi(a)}$ (a is any parameter) $\qquad$ *Rule ∼∃* $\quad \dfrac{\sim\exists x\varphi(x)}{\sim\varphi(a)}$ (a is any parameter)

Rule $\exists$ $\dfrac{\exists x\varphi(x)}{\varphi(a)}$ (a must be new) *Rule* $\sim\forall$ $\dfrac{\sim\forall x\varphi(x)}{\sim\varphi(a)}$ (a must be new)

Unified Notation

We recall the unifying α, β notation. We continue to use this as we did in Propositional Logic, except that "formula" will now mean closed formula of First-Order Logic (a formula is closed if it has no free variables; however, it may contain parameters). We now add two more categories γ and δ as follows:

For signed formulas, γ (read as "gam'-ma") shall be any formula of universal type, i.e. $T\ \forall x\varphi(x)$ or $F\ \exists x\varphi(x)$, and by $\gamma(a)$ we respectively mean $T\ \varphi(a)$, $F\ \varphi(a)$. δ (read as "del'ta") shall be any formula of existential type, that is $T\ \exists x\varphi(x)$ or $F\ \forall x\varphi(x)$, and by $\delta(a)$ we respectively mean $T\ \varphi(a)$, $F\ \varphi(a)$. Our universal rules $T\ \forall$ and $F\ \exists$ are now subsumed under Rule C below, and our existential rules $T\ \exists$ and $F\ \forall$ are subsumed under Rule D below:

Rule C. $\dfrac{\gamma}{\gamma(a)}$ *Rule D* $\dfrac{\delta}{\delta(a)}$ (providing a is new)

And, to review the propositional rules:

Rule A $\dfrac{\alpha}{\alpha_1}$ $\dfrac{\alpha}{\alpha_2}$ Rule B

$$\begin{array}{ccc} & \beta & \\ / & & \backslash \\ \beta_1 & & \beta_2 \end{array}$$

Thus, using our unifying notation, our twelve tableau rules for First-Order Logic are collapsed into four.

For *unsigned* formulas, we let γ be any formula of the form $\forall x\varphi(x)$ or $\sim\exists x\varphi(x)$, and by $\gamma(a)$ we mean $\varphi(a)$ or $\sim\varphi(a)$ respectively. We let δ be any formula of the form $\exists x\varphi(x)$ or $\sim\forall x\varphi(x)$, and by $\delta(a)$ we respectively mean $\varphi(a)$, $\sim\varphi(a)$.

Let us now try another tableau, namely let us prove the formula

$$\forall x\forall y(Px \supset Py) \supset (\forall x Px \vee \forall x{\sim}Px).$$

(1) $F\ \forall x\forall y(Px \supset Py) \supset (\forall x Px \vee \forall x{\sim}Px)$

(2) $T\ \forall x\forall y(Px \supset Py)$ By (1).

(3) $F\ \forall x Px \vee \forall x{\sim}Px$ By (1).

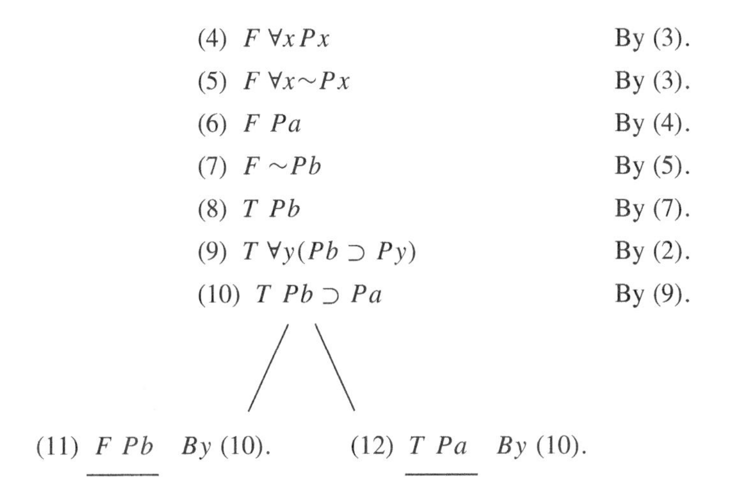

Discussion. In line (7), in accordance with Rule D, I couldn't use the parameter a a second time, so I had to take a new parameter, b. Now, in line (9), how did I know that I had best use the parameter b rather than a, or some other parameter? I knew because I informally went through a proof in my mind before I made the tableau, and then made the tableau accordingly.

Tableaux for Propositional Logic are purely routine things. It makes no essential difference in which order one uses the rules. If one order leads to a closure, so will any other order. But with first-order tableaux, the situation is entirely different. For one thing, in constructing a tableau for Propositional Logic, if one does not repeat any formula, the tableau must terminate in a finite number of steps, whereas in a first-order tableau, the process can go on indefinitely, because when we use a universal formula γ, we can add to our tableau $\gamma(a), \gamma(b), \ldots,$ with no limit to the number of parameters we can use. However, if one doesn't do things in the right order, the tableau might run on forever without ever closing, even though the tableau could be closed by proceeding in a different order. You might wonder if there could be some systematic procedure which, if followed, would guarantee closure if closure is possible. In fact there is, and we will consider one in the next section. Following the procedure is purely mechanical – it would be easy to program a computer to do it. However, a tableau constructed with intelligence and ingenuity usually closes far more quickly than one constructed using that purely mechanical procedure. We will return to this later.

Meanwhile, here are some strategic points that should be helpful: At any stage of the construction of a first-order tableau, it is wise to use all unused α's and β's before using γ's and δ's. Then use any available δ's (but not more than once, as I have already advised). As for the γ's, use any parameters already on the tree before introducing new ones.

In the exercises below, we recall that formulas of the form $X \equiv Y$ are treated thus:

$$\begin{array}{ccc} & T\ X \equiv Y & \\ & / \quad \backslash & \\ T\ X & & F\ X \\ T\ Y & & F\ Y \end{array} \qquad \begin{array}{ccc} & F\ X \equiv Y & \\ & / \quad \backslash & \\ T\ X & & F\ X \\ F\ Y & & T\ Y \end{array}$$

Also, in proving a formula of the form $X \equiv Y$, it reduces clutter to make two tableaux, one starting with $T\ X$ followed by $F\ Y$, and the other starting with $F\ X$ followed by $T\ Y$.

Exercise 1. Prove the following formulas using first-order tableaux:

(a) $\forall x(\forall y Py \supset Px)$
(b) $\forall x(Px \supset \exists x Px)$
(c) $\sim\exists y Py \supset \forall y(\exists x Px \supset Py)$
(d) $\exists x Px \supset \exists y Py$
(e) $(\forall x Px \wedge \forall x Qx) \equiv \forall x(Px \wedge Qx)$
(f) $(\forall x Px \vee \forall x Qx) \supset \forall x(Px \vee Qx)$
(g) $\exists x(Px \vee Qx) \equiv (\exists x Px \vee \exists x Qx)$
(h) $\exists x(Px \wedge Qx) \supset (\exists x Px \wedge \exists x Qx)$

Problem 1. The converse of (f) in the above exercise, i.e. the formula $\forall x(Px \vee Qx) \supset (\forall x Px \vee \forall x Qx)$, is not valid. Why? Also, the converse of (h) in the above exercise is not valid. Why?

Exercise 2. In this group of formulas to be proved by the tableau method, C is any closed formula (and thus for any parameter a, the formula $C(a)$ is simply C).

(a) $\forall x(Px \vee C) \equiv (\forall x Px \vee C)$
(b) $\exists x(Px \wedge C) \equiv (\exists x Px \wedge C)$
(c) $\exists x C \equiv C$
(d) $\forall x C \equiv C$
(e) $\exists x(C \supset Px) \equiv (C \supset \exists x Px)$
(f) $\exists x(Px \supset C) \equiv (\forall x Px \supset C)$
(g) $\forall x(C \supset Px) \equiv (C \supset \forall x Px)$
(h) $\forall x(Px \supset C) \equiv (\exists x Px \supset C)$
(i) $\forall x(Px \equiv C) \supset (\forall x Px \vee \forall x{\sim}Px)$

Tableau Completeness

Before proving that the tableau method for First-Order Logic is *complete*, i.e. that for every valid closed formula X there is a closed tableau for $F\ X$, we must make sure that the method is *correct*, i.e. that if there is a closed tableau for $F\ X$, then X really is valid. Equivalently, we must show that if $F\ X$ is satisfiable, then no tableau for $F\ X$ can close. It suffices to show that if a branch θ of a tableau is satisfiable, then any extension of θ by Rule A, Rule C or Rule D is again satisfiable, and that if θ is split into the two branches θ_1, θ_2 by Rule B, then at least one of the branches θ_1, θ_2 is satisfiable.

Thus, we must verify that for any satisfiable set S of formulas, the following facts hold:

F_1: For any α in S, the sets $S \cup \{\alpha_1\}$and $S \cup \{\alpha_2\}$ are both satisfiable.
F_2: For any β in S, either $S \cup \{\beta_1\}$ or $S \cup \{\beta_2\}$ is satisfiable.
F_3: For any γ in S, the set $S \cup \{\gamma(a)\}$ is satisfiable, where a is any parameter.
F_4: For any δ in S, the set $S \cup \{\delta(a)\}$ is satisfiable, providing a is a parameter not in any element of S.

Facts F_1, F_2, F_3 are obvious. Less obvious, though true, is F_4. In fact, it will be useful to observe the following stronger fact: We say that an interpretation I of a set S of closed formulas *satisfies* S if all elements of S are true under I. We say that a parameter a is *new* to S if it does not occur in any element of S. Now consider a set S of sentences and another set S' of sentences such that S is a subset of S'. Consider an interpretation I' of S' over the same domain U. We say that I' *extends* I, or is an *extension* of I, if for every predicate and parameter of (some formula of) S, its value under I is the same as its value under I'. [By its *value* under I is of course meant the element or relation of U assigned to it under I]. The following is a more explicit statement than F_4:

F_4*: If I satisfies S, then for any δ in S and any parameter a that is new to S, the set $S \cup \{\delta(a)\}$ is satisfied by some extension of I.

Problem 2. Prove F_4*.

Hintikka Sets

For First-Order Logic, we assume denumerably many parameters are available. We define a set S of signed sentences (sentences are closed formulas) to be a *Hintikka Set* for First-Order Logic if it satisfies the following conditions:

H_0: No formula and its conjugate are in S (as in Propositional Logic).
H_1: For any α in S, both α_1 and α_2 are in S (as in Propositional Logic).
H_2: For any β in S, either β_1 or β_2 is in S (as in Propositional Logic).
H_3: For any γ in S, the formula $\gamma(a)$ is in S for *every* parameter a.
H_4: For any δ in S, there is at least one parameter a such that the formula $\delta(a)$ is in S.

Hintikka's Lemma for First-Order Logic. Every Hintikka set for First-Order Logic is satisfiable in a denumerable domain.

The proof of the above lemma is not much more difficult than that for propositional logic.

Problem 3. Prove Hintikka's Lemma for First-Order Logic. Hint: Show that every Hintikka set is satisfiable in the denumerable domain of the parameters.

Note. We will occasionally be considering a *finite* domain D of parameters and a set S of closed formulas, all of whose parameters are in D. We define such a set to be a Hintikka set *for the domain D* as we did above, only in H_3 replacing "every parameter" by "every parameter in D." The proof given of Hintikka's Lemma for an infinite domain is easily modified to show that every Hintikka set for the finite domain D is satisfiable in the finite domain D.

Now we come to the completeness proof for first-order tableaux. This is more remarkable than that for Propositional Logic. In Propositional Logic a tableau for a formula must terminate after finitely many steps, and upon completion, the set of formulas on any open branch will be a Hintikka set (i.e. a Hintikka set for Propositional Logic). But a tableau for First-Order Logic may run on infinitely without ever closing, in which case there is at least one infinite branch θ (by König's Lemma), but the set of formulas on θ is not necessarily a Hintikka set! It may well be that some formula on the branch θ that could have been used with one of the rules has failed to be used. The point now is to devise some *systematic* procedure that will guarantee that if the tableau runs on infinitely, then for any open branch, the set of formulas on the branch *will* be a Hintikka set! There are many such procedures in the literature, and here is the one that I use.

For any non-atomic formula X on an open branch θ define X to be *fulfilled* on θ if:

(1) X is an α and α_1 and α_2 are both on θ, or
(2) X is a β and either β_1 or β_2 are on θ, or
(3) X is some γ and for every parameter a, the sentence $\gamma(a)$ is on θ, or
(4) X is some δ and $\delta(a)$ is on θ, for at least one parameter a.

To say that every point of an open infinite branch θ is fulfilled, is to say that the set of points on θ is a Hintikka set.

A Systematic Procedure for Generating a Tableau

Now, here is a systematic procedure that will ensure that if the tableau is infinite, then for any open branch θ, all the formulas on θ are fulfilled. In this procedure for generating a tableau, at each stage of the construction, certain points are declared to have been *used.* [As a practical bookkeeping device, we could put a check mark to the right of a formula as soon as it has been used.] We start the tableau by placing at the origin the formula whose satisfiability we are testing. This concludes the first stage. Now suppose we have completed the n-th stage. Our next act is determined as follows: If the tableau at hand is closed, we stop. If not, we pick an unused point X as high up in the tree as possible (say the leftmost one, if we want our procedure to be completely determinate). Then we take every open branch θ passing through X (there can only be finitely many such branches at any given point in the construction of the tableau) and proceed as follows:

(1a) If X is an α, β or δ, we apply Rule A, B or D respectively.

(1b) If X is some γ (and this is the delicate case!), we take the first parameter a (in some pre-arranged order of the parameters) for which the formula $X(a)$ does not already occur on θ and we extend θ to θ, $X(a)$, X; that is, we first add the new formula $X(a)$ as an end point to θ, getting θ, $X(a)$; but then we repeat the γ-formula X by adding it as an endpoint of the branch θ, $X(a)$.

(2) After applying either (1a) or (1b) to X, we then declare X to be used.

In this procedure, we are systematically working down the tree, fulfilling all α, β and δ formulas that come our way. As for the γ-formulas, when we use an occurrence of γ on a branch θ to subjoin an instance γ(a), the purpose of *repeating* an occurrence of γ is that we must sooner or later come down the branch θ, $\gamma(a)$, γ and be forced, by the rules of our systematic procedure, to use the repeated occurrence of γ, from which we adjoin another instance $\gamma(b)$ and again repeat an occurrence of γ, which in turn we later use again, and so forth. In this way we fulfill all the γ's too. Thus if the tableau runs on infinitely without closing, on any open branch, all the elements are fulfilled, and thus comprise a Hintikka set, which is simultaneously satisfiable in the denumerable domain of the parameters.

By *systematic tableau* we shall mean a tableau constructed by the above procedure. We thus see that is a systematic tableau doesn't close, then the origin is satisfiable, in fact in a denumerable domain. Therefore if the origin is not satisfiable, any *systematic* tableau must close. If now X is a valid closed formula, then $F\ X$ is not satisfiable, hence any systematic tableau for $F\ X$ must close, and X is thus provable by the tableau method. Also, if an unsigned formula X is satisfiable, so is the signed formula $T\ X$; hence no tableau for $T\ X$ can close, and so any *systematic* tableau for $T\ X$ runs on forever, has an infinite branch (by König's Lemma), and the set of elements of the branch is satisfiable in a denumerable domain. Thus $T\ X$ is satisfiable in that denumerable domain, and so is X.

We have thus killed two birds with one stone, having shown:

The Tableau Completeness Theorem. Every valid formula is provable by the tableau method.

Löwenheim Theorem. Every satisfiable formula is satisfiable in a denumerable domain.

Remarks

The reader should not be concerned by our apparently adding a new tableau rule, namely that of repeating the occurrence of a formula on a branch. This is just a book-keeping device for our systematic tableaux to force us to keep using each γ formula on every open branch below the γ, successively with every parameter in the system. We could just as easily imagine writing a note to ourselves (next to the $\gamma(a)$) at any place in the tableau where we conclude a particular $\gamma(a)$ from a particular γ, namely a note telling us to use that same γ again before applying the rules of the systematic procedure to the $\gamma(a)$ we just inferred. Of course alternatively we could just add repeating a formula to the tableau rules from the beginning, since that would do no harm.

As already mentioned, in general a systematic tableau is usually much longer than one constructed with ingenuity. Although a computer can be programmed to construct a systematic tableau, a clever human can usually construct a more efficient and shorter proof of a valid formula. As remarked by the logician Paul Rosenbloom in regards to a similar matter, "This proves that brains can sometimes be useful." If the reader has worked out some of the tableau exercises, it is highly unlikely that any of them were systematic. It would be a good exercise to re-do some of them with a systematic tableau and compare their length with the non-systematic one.

There are other systematic procedures in the literature, some of which yield quicker results that the one I have given, but those methods are more difficult to justify. Many improvements are possible. Such a study is a subject in itself and is known as "mechanical theorem proving."

Satisfiability in a Finite Domain

It can happen that in the construction of a tableau, a stage is reached in which the tableau is not closed, yet there is an open branch whose set of points is a Hintikka set *for the finite domain of parameters that occur on the branch.* In this case there is no point in continuing further, since we then know that the set of points on the branch (and this includes the origin) is satisfiable in that finite domain.

As an example, consider the formula $\forall x(Px \vee Qx) \supset (\forall x Px \vee \forall x Qx)$. This formula is not valid, hence the signed formula $F\ \forall x(Px \vee Qx) \supset (\forall x Px \vee \forall x Qx)$ is satisfiable. The following tableau reveals that the above signed formula is in fact satisfiable in a 2-element domain:

(1) $F\ \forall x(Px \vee Qx) \supset (\forall x Px \vee \forall x Qx)$

(2) $T\ \forall x(Px \vee Qx)$ By (1).

(3) $F\ \forall x Px \vee \forall x Qx$ By (1).

(4) $F\ \forall x Px$ By (3).

(5) $F\ \forall x Qx$ By (3).

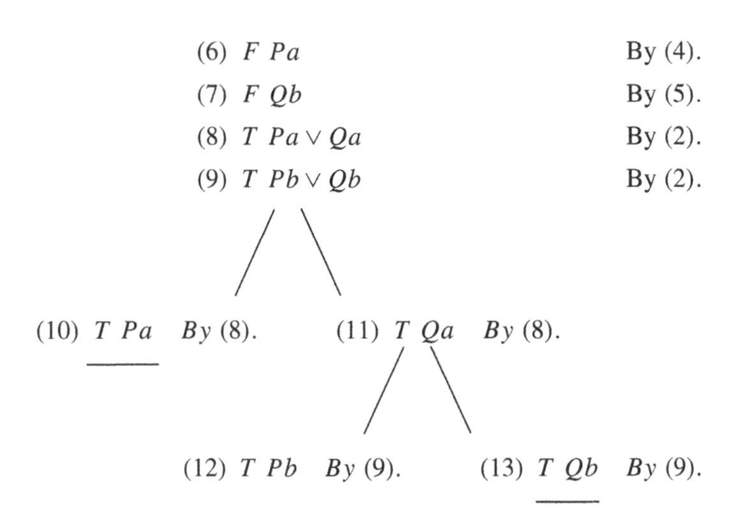

The set of formulas on the open branch ending with formula (12) is a Hintikka set for the 2-element domain $\{a, b\}$. We thus interpret Pb and Qa to be true, and Pa and Qb to be false, or, what is the same thing, we interpret P to be the set whose only element is b and Q as the set whose only element is a.

Discussion. We thus see that a first-order tableau can be used, not only to show certain formulas to be unsatisfiable, or equivalently, that certain formulas are valid, but also sometimes that certain formulas are satisfiable, if they happen to be satisfiable in a finite domain. The real mystery class consists of those formulas that are neither unsatisfiable nor satisfiable in a finite domain. If we construct a tableau for such a formula, it will run on forever, and at no stage will it either close nor reveal a finite Hintikka set.

The Skolem-Löwenheim Theorem and the Compactness Theorem

We have already stated that Löwenheim's theorem was extended by Skolem to the result that for any denumerable set S of sentences of First-Order Logic, if S is satisfiable at all, then it is satisfiable in a denumerable domain. There is also the *Compactness Theorem for First-Order Logic* – namely that for any denumerable set S of first-order sentences, if every finite subset of S is satisfiable, then the whole set S is (simultaneously) satisfiable. The following theorem combines these two results and yields each one:

Theorem L.S.C. [Löwenheim, Skolem, Compactness]. If S is a set of closed formulas without parameters, and all finite subsets of S are satisfiable, then the entire set S is satisfiable in a denumerable domain.

Of course the Löwenheim-Skolem Theorem follows from the above, because if a set S is satisfiable, all its finite subsets (in fact all its subsets) are obviously satisfiable.

There are several ways of proving the above theorem. One way is the following, which uses tableaux.

We are given a denumerable set S of closed formulas without parameters such that all finite subsets of S are satisfiable. As in the case of Propositional Logic, we arrange the elements of S in some denumerable sequence $X_1, X_2, \ldots, X_n, \ldots$. We start the tableau with X_1 at the origin. This concludes the first stage, and at any stage n, we proceed *systematically*, as already explained, and also append X_{n+1} at the end of every open branch. Since every finite subset of S is satisfiable, then at no stage can the tableau close. Hence it has an infinite open branch θ, and the set of formulas on θ is a Hintikka set (since the construction is systematic), and contains all the elements of S. Thus S is satisfiable in the denumerable domain of the parameters.

Boolean and First-Order Valuations

We recall the difference between a first-order formula being *valid* and being a *tautology*. Closely related to this is the following. We consider the set of all *sentences* – closed formulas with or without parameters. By a *valuation* v we shall mean an assignment of a truth value t or f to each sentence X. By $v(X)$ we mean the value (t or f) assigned to X under v. We say that X is *true* under v iff $v(X) = t$, and that X is *false* under v iff $v(X) = f$. A valuation v will be called a *Boolean valuation* if for all sentences X and Y the following four conditions hold:

B_1: $\sim X$ is true under v iff X is false under v.
B_2: $X \wedge Y$ is true under v iff both X and Y are true under v.
B_3: $X \vee Y$ is true under v iff at least one of X, Y is true under v.
B_4: $X \supset Y$ is true under v iff X is false under v or Y is true under v.

A Boolean valuation might also be called a valuation that respects the logical connectives.

We call v a *first order valuation* iff v is a Boolean valuation and also respects the quantifiers, in the sense that for any formula $\varphi(x)$, with x as the only free variable, the following two conditions hold:

- $\forall x\varphi(x)$ is true under v iff and only if for every parameter a, the sentence $\varphi(a)$ is true under v.
- $\exists x\varphi(x)$ is true under v iff and only if for at least one parameter a, the sentence $\varphi(a)$ is true under v.

To say that X is a *tautology* is to say that X is true under all *Boolean* valuations.

To say that X is *valid* (in the domain D of the parameters) is to say that X is true under all first-order valuations.

First-order valuations are very closely related to interpretations in the domain D of the parameters. Let us say that such an interpretation I *agrees* with v, or that v *agrees* with I, if for every sentence X, X is true under I iff X is true under v. Given any interpretation I, there is one and only one valuation v that agrees with I, namely that valuation that assigns t to those and only those sentences that are true under I. Also, for any valuation v, there is one and only one interpretation I that agrees with v, namely for any n-ary predicate P, take $I(P)$ to be the set of all n-tuples $(a_1, \ldots, a_n)$ such that $Pa_1, \ldots, a_n$ is true under v. [An n-ary relation can be regarded as a set of n-tuples.]

We shall say that a sentence X is *tautologically implied* by a set S if X is true under all *Boolean* valuations that satisfy S. For any S that is a finite set $\{X_1, \ldots, X_n\}$, this is equivalent to saying that $X_1 \wedge \ldots \wedge X_n \supset X$ is a tautology. We shall say that X is *validly implied* by S, if X is true under all *first-order* valuations that satisfy S.

The Regularity Theorem

We now turn to a basic result in First-Order Logic, which, among other things, provides a particularly neat proof of the completeness of the axiom system of the last chapter, as well as for related axiom systems.

We work now with unsigned formulas. We must first define the concept of a *regular formula*, of which there are two types. By a *regular formula* of type C we shall mean a formula of the form $\gamma \supset \gamma(a)$. By a *regular formula* of type D we shall mean a formula of the form $\delta \supset \delta(a)$, where a is a parameter that does not occur in δ. We shall use the letter Q to denote a regular formula; thus Q will mean either γ or δ, and $Q(a)$ will then mean $\gamma(a)$ or $\delta(a)$, respectively. By a *regular sequence* we shall mean a finite (possibly empty) sequence $Q_1 \supset Q_1(a_1), \ldots, Q_n \supset Q_n(a_n)$. such that each term is regular, and furthermore, for each $i < n$, if Q_{i+1} is a δ, then a_{i+1} does not occur in any of the earlier terms $Q_1 \supset Q_1(a_1), \ldots, Q_i \supset Q_i(a_i)$. By a *regular set* R, we shall mean a finite set of formulas whose members can be arranged in a regular sequence. Alternatively, a regular set can be characterized as any finite set constructed according to the following rules:

R_0: The empty set $\emptyset$ is regular.
R_1: If R is regular, so is $R \cup \{\gamma \supset \gamma(a)\}$.
R_2: If R is regular, so is $R \cup \{\delta \supset \delta(a)\}$, providing a does not occur in δ or in any element of R.

We shall call the parameter a a *critical parameter* of a regular set R if for any δ, the sentence $\delta \supset \delta(a)$ is a member of R. Our first aim is to show that if X is validly implied by a regular set R, and no critical parameter of R occurs in X, then X is valid. After that, we will show the remarkable result that every valid sentence X

is *tautologically* implied by some regular set R – in fact one such that no critical parameter of R occurs in X.

Problem 4. (a) Prove that if the set S is satisfiable and X is valid, then $S \cup \{X\}$ is satisfiable. [Pretty obvious, eh?]
(b) Prove that if S is satisfiable and $S \cup \{X\}$ is not satisfiable, then for any sentence Y, the set $S \cup \{X \supset Y\}$ is satisfiable.

Problem 5. Suppose S is a finite set of (simultaneously) satisfiable sentences. Prove:

(a) For every parameter a, the set $S \cup \{\gamma \supset \gamma(a)\}$ is satisfiable.
(b) For any parameter a that occurs neither in δ nor in any element of set S, the set $S \cup \{\delta \supset \delta(a)\}$ is satisfiable. [Hint: Consider the two possible cases: $S \cup \{\delta\}$ is satisfiable and $S \cup \{\delta\}$ is not satisfiable.]
(c) If S is satisfiable and R is a regular set such that no critical parameter of R occurs in S, then $R \cup S$ is satisfiable.
(d) Every regular set is satisfiable.
(e) If X is validly implied by a regular set R, and if no critical parameter of R occurs in X, then X is valid.
(f) If $(\gamma \supset \gamma(a)) \supset X$ is valid, so is X.
(g) If $\delta \supset \delta(a)$ is regular and validly implies X, and if no critical parameters of $\delta \supset \delta(a)$ occurs in X, then X is valid.

Discussion. Concerning (d) of the above problem, a regular set R is not only satisfiable, but has an even stronger property that is midway in strength between satisfiability and validity: Consider first a single sentence $\varphi(a_1, \ldots, a_n)$ whose parameters are $a_1, \ldots, a_n$. We shall call the sentence *strongly satisfiable* if for every interpretation I of the predicates of the formula in some universe U, there are elements $e_1, \ldots, e_n$ of U such that $\varphi(e_1, \ldots, e_n)$ is true under I. [Note: This condition is equivalent to the following: Let $x_1, \ldots, x_n$ be variables that do not occur in the sentence $\varphi(a_1, \ldots, a_n)$, and let $\varphi(x_1, \ldots, x_n)$ be the result of substituting $x_1, \ldots, x_n$ for the parameters $a_1, \ldots, a_n$, respectively. Then $\varphi(a_1, \ldots, a_n)$ is strongly satisfiable if and only if the sentence $\exists x_1 \exists x_2 \ldots \exists x_n \varphi(x_1, \ldots, x_n)$ is valid.] Now consider a set S of sentences. We shall call S *strongly satisfiable* if for every interpretation I of the predicates of the formulas of S, there is a choice of values for the parameters of the formulas of S that makes all the elements of S true. Well, a regular set R is not only satisfiable, but even strongly satisfiable – in fact, it has the even stronger property that for every interpretation of the predicates of R and any choice of values for the non-critical parameters of R, there is a choice of values for the critical parameters of R that makes all the elements of R true. We leave the verification of this as an exercise for the reader.

Now for the main result:

Theorem R [Regularity Theorem]. Every valid sentence X is tautologically implied by some regular set R such that no critical parameter of R occurs in X.

We will prove this theorem by showing how, from a closed tableau $\mathcal{T}$ for $\sim X$, we can find such a set R. The method is beautifully simple! Just take for R the set of all formulas $Q \supset Q(a)$ such that $Q(a)$ was inferred from Q by Rule C or Rule D! We will call this set the *associated regular set* of $\mathcal{T}$.

Problem 6. Prove that the set R works. [Hint: Construct another tableau $\mathcal{T}_1$ starting with $\sim X$ and the elements of R arranged in a regular sequence. Show that $\mathcal{T}_1$ can be made to close using only Rules A and B.]

Let us consider an example. Let X be $\forall x(Px \supset Qx) \supset (\exists x Px \supset \exists x Qx)$. Here is a closed tableau for the sentence $\sim X$ (just for a change of pace, we use unsigned formulas in this tableau):

(1) $\sim(\forall x(Px \supset Qx) \supset (\exists x Px \supset \exists x Qx))$

(2) $\forall x(Px \supset Qx)$ By (1).

(3) $\sim(\exists x Px \supset \exists x Qx)$ By (1).

(4) $\exists x Px$ By (3).

(5) $\sim\exists x Qx$ By (3).

(6) Pa By (4).

(7) $\sim Qa$ By (5).

(8) $Pa \supset Qa$ By (2).

(9) $\sim Pa$ *By* (8). (10) Qa *By* (8).

We have inferred (6) from (4) by Rule D, (7) from (5) by Rule C, and (8) from (2) by Rule C. Thus our regular set is $\{(4) \supset (6), (5) \supset (7), (2) \supset (8)\}$, that is, $\{\exists x Px \supset Pa, \sim\exists x Qx \supset \sim Qa, \forall x(Px \supset Qx) \supset (Pa \supset Qa)\}$. To see more clearly that X is tautologically implied by R, let us use the following abbreviations:

$$p = Pa$$
$$q = Qa$$
$$r = \exists x Px$$
$$s = \exists x Qx$$
$$m = \forall x(Px \supset Qx)$$

Then R is the set $\{r \supset p, \sim s \supset \sim q, m \supset (p \supset q)\}$. One can then see that X is tautologically implied by R, in other words, that

$$[(r \supset p) \wedge (\sim s \supset \sim q) \wedge (m \supset (p \supset q))] \supset (m \supset (r \supset s))$$

is a tautology.

One can see this, using a truth table, or more simply by a tableau using only Rules A and B.

One obtains some very curious tautologies in this manner. It should be fun for the reader to take some of the valid formulas proved in earlier exercises and find the regular sets that tautologically imply them.

Completeness of the Axiom System $\mathcal{S}_1$

We recall the axiom system $\mathcal{S}_1$ for First-Order Logic of the last chapter:

Axioms

Group 1. All tautologies.

Group 2. (a) All sentences $\forall x\varphi(x) \supset \varphi(a)$.

(b) All sentences $\varphi(a) \supset \exists x\varphi(x)$.

Inference Rules.

I. Modus Ponens $\dfrac{X,\ X \supset Y}{Y}$.

II. (a) $\dfrac{\varphi(a) \supset X}{\exists x\varphi(x) \supset X}$, (b) $\dfrac{X \supset \varphi(a)}{X \supset \forall x\varphi(x)}$,

where X is closed and a is a parameter that does not occur in either X or $\varphi(x)$.

We have already proved that the system $\mathcal{S}_1$ is correct (everything provable is valid). We now wish to prove the system complete (all valid sentences are provable).

The regularity theorem provides a neat proof of the completeness of $\mathcal{S}_1$, as we will see.

First, consider an arbitrary axiom system $\mathcal{A}$ for First-Order Logic. We shall say that $\mathcal{A}$ is *tautologically closed* if all tautologies are provable in $\mathcal{A}$, and for any finite set S of formulas provable in $\mathcal{A}$, any formula X that is tautologically implied by S is also provable in $\mathcal{A}$.

Theorem 1. If $\mathcal{A}$ is tautologically closed and obeys conditions (A_1) and (A_2) below, then $\mathcal{A}$ is complete.

(A_1) If $(\gamma \supset \gamma(a)) \supset X$ is provable in $\mathcal{A}$, so is X.

(A_2) If $(\delta \supset \delta(a)) \supset X$ is provable in $\mathcal{A}$ and a does not occur in either δ or X, then X is provable in $\mathcal{A}$.

This theorem follows rather easily from the Regularity Theorem.

Problem 7. Prove the above theorem.

Next, the following lemma about the system $\mathcal{S}_1$ will be helpful.

Lemma. (a) For any γ, the sentence $\gamma \supset \gamma(a)$ is provable in $\mathcal{S}_1$. (b) For any δ, if $\delta(a) \supset X$ is provable in $\mathcal{S}_1$ and a does not occur in δ or in X, then $\delta \supset X$ is provable in $\mathcal{S}_1$.

Problem 8. First prove the above lemma. Then prove the completeness of $\mathcal{S}_1$ by showing that $\mathcal{S}_1$ satisfies the conditions of Theorem 1.

Remark. Theorem 1 actually provides another complete axiom system in its own right, namely one may just take as axioms all tautologies, and for inference rules take modus ponens and the following two:

$$R_1\colon \frac{(\gamma \supset \gamma(a)) \supset X}{X}$$

$$R_2\colon \frac{(\delta \supset \delta(a)) \supset X}{X}, \text{ providing } a \text{ does not occur in } \delta \text{ or in } X$$

Solutions to the Problems of Chapter 9

1. (a) The converse of (f) is $\forall x(Px \vee Qx) \supset (\forall x Px \vee \forall x Qx)$ To show that this converse is not valid, it suffices to show that it is false under at least one interpretation. Well, consider the set of all (natural) numbers, and interpret P to be the set of all even numbers and Q to be the set of all odd numbers. Then $\forall x(Px \vee Qx)$ is true (every number is either even or odd), but $\forall x Px$ and $\forall x Qx$ are both false (it is false that all numbers are even, and it is false that all numbers are odd). Hence $\forall x Px \vee \forall x Qx$ is false, and since $\forall x(Px \vee Qx)$ is true, then entire implication $\forall x(Px \vee Qx) \supset (\forall x Px \vee \forall x Qx)$ is false.

   (b) The converse of (h) is $(\exists x Px \wedge \exists x Qx) \supset \exists x(Px \wedge Qx)$. Consider the same interpretation as in (a). Then $\exists x(Px \wedge Qx)$ is true (there is an even number and there is an odd number), but $\exists x(Px \wedge Qx)$ is obviously false (it is false that there exists a number which is both even and odd). Hence $(\exists x Px \wedge \exists x Qx) \supset \exists x(Px \wedge Qx)$ is false. Thus the formula is not valid.
2. We are given that I is an interpretation of all the predicates and parameters of the formulas of the set S in a non-empty domain U in which all formulas of S are true, and that δ is an element of S and a is a parameter new to S. Since a is new to S, then I has not assigned any element of U to the parameter a. Since δ is in S, then δ is true for some element e of U. Since a has not been assigned any value in U, we now extend the interpretation I by assigning to a the element e. Thus

$\delta(a)$ is true under the extension of I. Hence all elements of the set of formulas $S \cup \{\delta(a)\}$ are true under the extension.

3. Let S be a Hintikka set of signed formulas. For each predicate P of degree n, we interpret P to be that relation $R(a_1, \ldots, a_n)$ on the set of parameters such that $T\ Pa_1, \ldots,\ a_n$ is an element of S. We now show by induction on the degree of X (i.e. by induction on the number of occurrences of logical connectives and quantifiers in X) that every X in S is true under the interpretation.

   Obviously every atomic element of S is true under the interpretation. Now suppose that X is of degree n and that all elements of S of degree less than n are true. We are to show that X is true.

   If X is an α or β, the proof is the same as that of Propositional Logic. Suppose X is some γ. Then $\gamma(a)$ is in S for all parameters a (by H_3). Since each $\gamma(a)$ is of degree less than n, it is true (by the induction hypothesis). Hence γ is true.

   If X is some δ, then for some parameter a, the sentence $\delta(a)$ is in S (by H_4), and being of lower degree than n, is true under the interpretation. This completes the induction.

4. (a) If S is satisfiable, then all elements of S are true under some interpretation I. If X is valid, then X is true under all interpretations, hence also true under I, and so all elements of $S \cup \{X\}$ are true under I.

   (b) Suppose S is satisfiable and $S \cup \{X\}$ is not. Then there is some interpretation I in which all elements of S are true, but X is not true (since X is not satisfiable). Since X is false under I, then for any sentence Y, the sentence $X \supset Y$ is true under I, and so all elements of $S \cup \{X \supset Y\}$ are true under I.

5. We are given that S is a finite set of simultaneously satisfiable sentences.
   (a) Since S is satisfiable and $\gamma \supset \gamma(a)$ is valid, then $S \cup \{\gamma \supset \gamma(a)\}$ is satisfiable by Problem 4 (a).
   (b) Since S is satisfiable, then if $S \cup \{\delta\}$ is not satisfiable, then $S \cup \{\delta \supset \delta(a)\}$ is satisfiable, by Problem 4 (b). On the other hand, if $S \cup \{\delta\}$ is satisfiable, so is $S \cup \{\delta\} \cup \{\delta(a)\}$ [a is new to $S \cup \{\delta\}$] by F_4, and $S \cup \{\delta\} \cup \{\delta(a)\}$ is the set $S \cup \{\delta, \delta(a)\}$. Since all elements of $S \cup \{\delta, \delta(a)\}$ are true under some interpretation, all elements of $S \cup \{\delta \supset \delta(a)\}$ will be true under the same interpretation.
   (c) Suppose S is satisfiable and R is regular and no critical parameter of R occurs in any element of S, Arrange R is some regular sequence $(r_1, r_2, \ldots r_n)$. If we successively adjoin $r_1, r_2, \ldots r_n$ to S, at no stage do we destroy satisfiability (by (a) and (b)). Hence the resulting set $R \cup S$ is satisfiable.
   (d) This follows from (c) by taking S to be the empty set.
   (e) Suppose that X is validly implied by some regular set R, and that no critical parameter of R occur in X. If X were not valid, then $\sim X$ would be satisfiable, hence $R \cup \sim X$ would be satisfiable (by (c)). Hence R would not validly

imply X, contrary to the given condition that R *does* validly imply X. Hence $\sim X$ cannot be satisfiable, which means that X is valid.

(f) If $(\gamma \supset \gamma(a)) \supset X$ is valid, then X must be valid, because $\gamma \supset \gamma(a)$ is itself valid.

(g) This follows from (e), by taking for R the regular set $\{\delta \supset \delta(a)\}$.

6. In constructing $\mathcal{T}_1$, we start the tableau with $F\ X$ and the elements of R, and then follow the construction of $\mathcal{T}$, but when we come to some Q on a branch θ, instead of directly inferring $Q(a)$ from Q, as we in on $\mathcal{T}$, we split θ to the two branches $(\theta, \sim Q)$ and $(\theta, Q(a))$, which we can do by Rule B, since $Q \supset Q(a)$ is above Q on the branch θ.

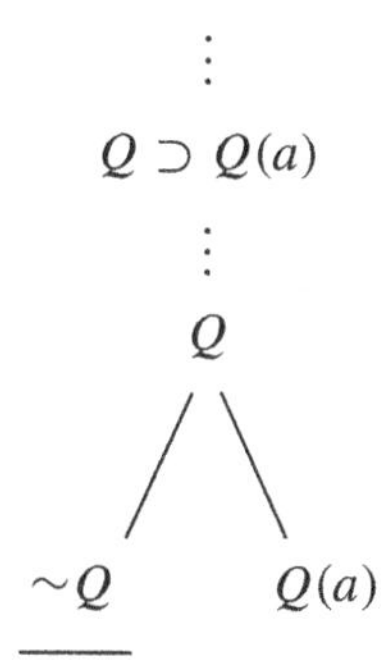

The left branch closes, since Q and $\sim Q$ are on it, and so we have effectively gone from Q to $Q(a)$ without using Rule C or Rule D. Thus $\mathcal{T}_1$ uses only Rule A and Rule B.

7. Suppose X is valid. By the Regularity Theorem, X is tautologically implied by some regular set R such that no critical parameter of R occurs in X. Arrange the elements of R in an *inverse* regular sequence, i.e. in a sequence $(r_1, \ldots, r_n)$ such that the sequence $(r_n, \ldots, r_1)$ is a regular sequence. Thus for each $i \leq n$, r_i is of the form $Q_i \supset Q_i(a_i)$, and if Q_i is a δ, then the critical parameter a_i does not appear in any later term of the sequence. Since X is tautologically implied by R, then the formula $(r_1 \wedge \ldots \wedge r_n) \supset X$ is a tautology, hence so is the formula $(r_1 \supset (r_2 \wedge \ldots \wedge r_n) \supset X)$, which is tautologically implied by it, and hence provable in $\mathcal{Q}$. Therefore $(r_2 \wedge \ldots \wedge r_n) \supset X$ is provable in $\mathcal{Q}$ [by A_1 or A_2, since no critical parameter of r_1 occurs in $(r_2 \wedge \ldots \wedge r_n) \supset X$]. If $n > 2$, then similarly $(r_2 \supset (r_3 \wedge \ldots \wedge r_n) \supset X)$, and then $(r_3 \wedge \ldots \wedge r_n) \supset X$ is provable in $\mathcal{Q}$. In this way, we successively eliminate $r_1, \ldots, r_n$ and obtain a proof of X.

8. First to prove the lemma:

(a) To show that $\gamma \supset \gamma(a)$ is provable in $\mathcal{S}_1$, we note that γ is either of the form $\forall x\varphi(x)$ or $\sim\exists x\varphi(x)$. If the former, then $\gamma \supset \gamma(a)$ is the sentence $\forall x\varphi(x) \supset \varphi(a)$, which is an axiom of $\mathcal{S}_1$. If the latter, then $\gamma \supset \gamma(a)$ is the sentence $\sim\exists x\varphi(x) \supset \sim\varphi(a)$, which is tautologically implied by the axiom $\varphi(a) \supset \exists x\varphi(x)$, hence is provable in $\mathcal{S}_1$.

(b) Suppose $\delta(a) \supset X$ is provable in $\mathcal{S}_1$ and a does not occur in δ or in X. If δ is of the form $\exists x\varphi(x)$, then $\delta(a) \supset X$ is the sentence $\varphi(a) \supset X$, and since it is provable in $\mathcal{S}_1$, so is $\exists x\varphi(x) \supset X$, by Rule II (a). Now suppose δ is of the form $\sim\forall x\varphi(x)$. Then $\delta(a) \supset X$ is the sentence $\sim\varphi(a) \supset X$, which is provable in $\mathcal{S}_1$, hence so is $\sim X \supset \varphi(a)$, which is tautologically implied by it. Then by Rule II (b), taking $\sim X$ for X, the formula $\sim X \supset \forall x\varphi(x)$ is provable, hence so is $\sim\forall x\varphi(x) \supset X$, which is tautologically implied by $\sim X \supset \forall x\varphi(x)$. Thus $\sim\forall x\varphi(x) \supset X$ is provable in $\mathcal{S}_1$, and this is the sentence $\delta \supset X$.

This proves the lemma. Now we must show that $\mathcal{S}_1$ satisfies the hypotheses of Theorem 1.

Since all tautologies are provable in $\mathcal{S}_1$ and modus ponens is a rule of $\mathcal{S}_1$, then of course $\mathcal{S}_1$ is closed under tautological implication. It remains to show that conditions (A_1) and (A_2) of the hypotheses of Theorem 1 hold for $\mathcal{S}_1$.

As for (A_1), suppose $(\gamma \supset \gamma(a)) \supset X$ is provable in $\mathcal{S}_1$. Also, $\gamma \supset \gamma(a)$ is provable in $\mathcal{S}_1$ (by the lemma). Hence X is provable in $\mathcal{S}_1$ (by modus ponens).

As for (A_2), suppose $(\delta \supset \delta(a)) \supset X$ is provable in $\mathcal{S}_1$ and a does not occur in either δ or X. The two formulas $\sim\delta \supset X$ and $\delta(a) \supset X$ are both tautologically implied by $(\delta \supset \delta(a)) \supset X$ [as the reader can verify], hence are both provable in $\mathcal{S}_1$. Since $\delta(a) \supset X$ is provable, and a does not occur in δ or X, then $\delta \supset X$ is provable (by (b) of the lemma). Thus $\delta \supset X$ and $\sim\delta \supset X$ are both provable, hence so is X, since it is tautologically implied by the two of them.

Part IV
The Incompleteness Phenomenon

10

Incompleteness in a General Setting

In the first third of the 20th century, there were two mathematical systems in existence that were so extensive that it was generally assumed that every mathematical proposition could be either proved or disproved within the systems. But soon the great logician Kurt Gödel amazed the entire mathematical world with a paper [1931] that showed that it was not the case that every mathematical proposition could be either proved or disproved within these systems. This famous result soon became known around the world as *Gödel's Incompleteness Theorem* (or sometimes just *Gödel's Theorem* for short). Gödel's paper begins with the following startling words:

> "The development of mathematics in the direction of greater precision has led to large areas of it being formalized, so that proofs can be carried out according to a few mechanical rules. The most comprehensive formal systems to date are, on the one hand, the *Principia Mathematica* of Whitehead and Russell, and, on the other hand, the Zermelo-Fraenkel system of axiomatic set theory. Both systems are so extensive that all methods of proof used in mathematics today can be formalized in them, i.e. can be reduced to axioms and rules of inference. It would seem reasonable, therefore, to surmise that these axioms and rules of inference are sufficient to decide all mathematical questions that can be formulated in the systems concerned. In what follows it will be shown that this is not the case, but rather that, in both of the cited systems, there exist relatively simple problems of the theory of ordinary whole numbers which cannot be decided on the basis of the axioms."

Gödel then goes on to explain that the theorems he will prove do not depend on the special nature of the two systems under consideration but rather hold for an extensive class of mathematical systems.

I will be considering systems using First-Order Logic here. In this chapter, the results will be quite general, and of the following form: "If a system has such-and-such features, then so-and-so follows." In the next chapter, we will consider an actual well-known system which in fact does have the "such-and-such" features. But before all this, I would like to illustrate the essential idea behind Gödel's proof with what may aptly be called (and was so called by the logician Henk Barendregt) a "mini Gödel theorem."

A Gödelian Machine

Let us consider a computing machine that prints out various expressions composed of the following five symbols: $\sim$ P N ( ).

An expression X is called *printable* if the machine can print it. We assume the machine is so programmed that any expression the machine can print will be printed sooner or later.

By the *norm* of an expression X, we mean the expression $X(X)$, e.g. the norm of $P{\sim}$ is $P{\sim}(P{\sim})$. By a *sentence* we mean any expression of one of the following four forms (where X is any expression whatever).

(1) $P(X)$
(2) ${\sim}P(X)$
(3) $PN(X)$
(4) ${\sim}PN(X)$

Informally, P stands for "printable", $\sim$ stands for "not", and N stands for "the norm of". And so we define $P(X)$ to be *true* iff X is printable, ${\sim}P(X)$ to be true iff X is not printable, $PN(X)$ to be true iff the norm of X is printable, and ${\sim}PN(X)$ to be true iff the norm of X is not printable. [We read "${\sim}PN(X)$" as "not printable the norm of X", or, in better English, "the norm of X is not printable".]

We have now given a perfectly precise definition of what it means for a sentence to be true. We now assume that we are also given that the machine is completely accurate in that all sentences printed by the machine are true. It never prints any false sentences. Thus, for example, if the machine ever prints $P(X)$, then X really is printable, and so the machine will print X sooner or later.

What about the converse? If X is printable, does it follow that $P(X)$ is also printable? Not necessarily. If X is printable, then $P(X)$ is certainly *true*, but that does not mean that $P(X)$ must be printable. We are given that all printable sentences are true, but we are not given that all true sentences are printable! As a matter of fact, there is a sentence which is true but not printable!!

Problem 1. Exhibit a true sentence which the machine cannot print. [Hint: Construct a sentence that asserts its own non-printability, that is, a sentence X that is true if and only if it is not printable.]

I must warn you that our arguments now are going to become a little more complex. It's true that in previous chapters, in order to really understand all the material and how it all fits together, it is likely to have been useful to read the whole chapter more than once. But in the coming pages, you may also have to think about individual definitions, proofs and problem solutions for a little more time than in the past when you first encounter them, and also reread chapters ... and sometimes even reread a few chapters together to see more clearly how the results of later chapters depend upon results in earlier chapters. From a mathematical logician's point of view what you have from here on to the end of the book is really is something like one long argument (based on many key definitions) establishing many inter-related (and famous) results. But if you manage working this through the material this way, you will have a very, very good understanding of the basics of modern mathematical logic when you are done.

Some Basic General Results

We shall begin by considering a very basic kind of system in which certain expressions are called *designators* (these were called *class-names* by Gödel), certain expressions are called *sentences*, and certain sentences are classified as *true sentences*, the remaining sentences being classified as *false sentences*. In what follows, the word *number* will mean *natural number* (zero or a positive integer).

Each designator H designates, or is a name of, a set of (natural) numbers. A set of numbers will be called *nameable* if some designator names it. To each designator H and each number n is assigned a sentence denoted by $H(n)$, which we call *true* if and only if n belongs to the set named by H. We shall sometimes refer to $H(n)$ as the result of *applying* H to n.

Remarks. In systems couched in the language of First-Order Logic, which we will soon consider, the "designators" are formulas with just one free variable. How $H(n)$ is to be defined will be considered later. There are two well-known ways of doing this, but for now we wish our results to be quite general, and so we will not yet specify how $H(n)$ comes about.

Gödel Numbering

Following Gödel, to each expression is assigned a number now called the *Gödel number* of the expression. Distinct expressions have distinct Gödel numbers. If n is the Gödel number of a sentence, then we will call n a *sentence number*, and we let S_n denote that sentence. If n is the Gödel number of a designator, we will call n a *designator number*, and we let H_n be the designator whose Gödel number is n. The sentence $H_n(n)$ (which is H_n applied to its own Gödel number) is called the *diagonalization* of H_n.

To each number n is assigned a (sentence) number denoted by n^* such that if n is a designator number, then n^* is the Gödel number of the diagonalization $H_n(n)$ of H_n.

Two sentences X and Y will be called *semantically equivalent* if they are either both true or both not true.

The systems under consideration obey the following two conditions:

C_1: To each designator H is assigned a designator K, called the *diagonalizer* of H, such that for any designator number n, the sentence $K(n)$ is semantically equivalent to $H(n^*)$.

C_2: To each designator H is assigned a designator H', called the *negation* of H, such that for any number n, the sentence $H'(n)$ is true if and only if $H(n)$ is not true.

Note: In systems couched in the language of First-Order Logic, a designator H is a formula $F(x)$ with one free variable x, and H' is then $\sim F(x)$.

From just those two conditions, something surprising follows (at least I hope the reader will be surprised!), namely:

Theorem T. [Tarski's Theorem in miniature]. The set of Gödel numbers of the true sentences is not nameable.

To prove Theorem T, we define a sentence S_n to be a *Gödel sentence* for a set A of numbers, if it is the case that S_n is true iff $n \in A$. Thus a Gödel sentence for A is a sentence which is true if and only if its Gödel number is in A, i.e. the Gödel sentences for a set A are all the sentences S_n (each with its sentence number n) such that either $n \in A$ and S_n is true or $n \notin A$ and S_n is not true. [Note that every sentence is either true or false. It is only designators that are true for some natural numbers and false for others.]

Lemma 1. For any nameable set A, there is a Gödel sentence for A.

The following definitions will be helpful in proving Lemma 1 and Theorem T: For any number set A, we define $A^{\#}$ to be the set of all n such that $n^* \in A$. Also, for any set of expressions A, we define A_0 to be the set of Gödel numbers of the expressions in A. From now on we will use the notation $\widetilde{A}$ for the complement of the set A. For example, we will consistently use the notation T_0 for the set of Gödel numbers of all *true* sentences of the system under consideration. Thus $\widetilde{T_0}$ will be the set of all natural numbers that are not Gödel numbers of a true sentence.

Problem 2. Prove Theorem T by following these three steps (the methods we will use in the solution set to solve these problems here will be equally important later):

(a) Prove that if A is nameable, so is $A^{\#}$. [Hint. Show that if A is named by the designator H, then $A^{\#}$ is named by the diagonalizer of H.]
(b) Now prove Lemma 1. [Hint. Show that if k is the Gödel number of a designator that designates $A^{\#}$, then $H_k(k)$ is a Gödel sentence for A.]
(c) Now prove Theorem T. [Hint. Show that $\widetilde{T_0}$ has no Gödel sentence.]

Enter Gödel

We continue to consider the system so far described, which satisfies conditions C_1 and C_2. But now let us assume that in addition the system contains a set of axioms and rules that can be used to prove various sentences. We further assume that the axioms and rules are correct in that only true sentences are provable. Now suppose that the axiom system is such that the set of Gödel numbers of the provable sentences *is* nameable. Let P be the set of provable sentences and let P_0 be the set of their Gödel numbers. Since P_0 is nameable and T_0 is not (by Theorem T), then the two sets don't coincide. Hence the set P and T don't coincide, which means that either some sentence X is in P but not in T or is in T but not in P; in other words X is provable but not true, or is true but not provable. The first alternative is ruled out by the given condition that only true sentences are provable. Hence X must be a sentence which is true but not provable from the axioms.

More can be said: Since P_0 is nameable, so is its complement $\widetilde{P_0}$ (namely by H', where H names P_0). Any Gödel sentence for the set $\widetilde{P_0}$ must be a sentence that is true but not provable from the axioms (why?). Hence the diagonalization of the diagonalizer of H' is such a sentence (why?).

Syntactic Incompleteness Theorems

The incompleteness proof that we just considered departs from what Gödel actually did in one important respect – namely that it involves the notion of *truth*, and accordingly would be called *semantical*. Gödel's original proof made no reference to the notion of truth or interpretation, but was involved only with the notion of formal provability, and accordingly would be classified as *syntactical*. We shall now follow him more closely.

We consider a type of system which, like the former, has certain expressions called *sentences* and other expressions that we will call *formulas* instead of *designators*. Again, to each formula F and each (natural) number n is assigned a sentence that we will denote $F(n)$. Also, to each sentence X is assigned a sentence denoted $\sim X$ that is called the *negation* of X, and to each formula F is assigned a formula F' such that for any number n, the sentence $F'(n)$ is the negation $\sim F(n)$ of $F(n)$. [In application to systems couched in the language of First-Order Logic, the "formulas" are first-order formulas with just one free variable, and F' is simply $\sim F$.]

We now have no notion of *true sentences*, but we have a well-defined set P of sentences called *provable* sentences. A sentence X is called *refutable* if its negation $\sim X$ is provable. We let R be the set of all *refutable* sentences. We call a sentence X *decidable* if either it or its negation is provable – in other words, X is decidable if it is either provable or refutable and *undecidable* if it is neither provable nor refutable. The system is called *complete* if every sentence is either provable or refutable, and *incomplete* otherwise. We call the system *consistent* if no sentence is both provable and refutable, and *inconsistent* if some sentence is both provable and refutable.

Again, each expression is assigned a Gödel number, and again, if n is the Gödel number of a sentence, then n is called a *sentence number*, and we let S_n be the sentence having n as its Gödel number. If n is the Gödel number of a formula, we left F_n be that formula, and we call n a *formula number*.

For any set W of expressions, we let W^* be the set of all numbers n such that $n^* \in W_0$. Thus $W^* = W_0{}^{\#}$.

The sets P^* and R^* play key roles in what follows.

Problem 3. Prove that if n is a formula number, then $n \in W^*$ iff $F_n(n) \in W$. [Thus for any formula number n, $n \in P^*$ iff $F_n(n)$ is provable, and $n \in R^*$ iff $F_n(n)$ is refutable.]

We say that a formula F *represents* the set of all numbers n such that $F(n)$ is provable. Thus, to say that F represents the number set A is to say that for all numbers n, the sentence $F(n)$ is provable iff $n \in A$.

Theorem G_0 [A forerunner of Gödel's Theorem]. If P^* is representable and the system is consistent, then some sentence is undecidable.

Problem 4. Proven Theorem G_0.

A variant. In my book *Theory of Formal Systems*[1961], I stated and proved the following variant of Theorem G_0.

Theorem S_0. If R^* is representable and the system is consistent, then some sentence is undecidable.

Problem 5. Prove Theorem S_0.

Discussion. The sentence we get from representing P^* is essentially Gödel's sentence, which asserts its own non-provability. It can be thought of as saying "I am not provable". By contrast, the sentence we get by representing R^* can be thought of as saying "I am refutable". This sentence was independently rediscovered by the logician R. G. Jerislow [1973], and used by him to accomplish certain things that could not be done with the Gödel sentence.

Problem 6. We have seen that in a consistent system, if F represents P^*, then $F(f')$ is undecidable, where f' is the Gödel number of F'. Now, suppose instead that F' represents P^*. Is $F(f)$ undecidable, for f the Gödel number of F?

Enter Rosser

In Gödel's proof, Gödel did indeed construct a formula that represented P^*, but to prove that it represented P^* in the systems he dealt with, he had to make an assumption (a perfectly reasonable one, in fact) that was stronger than consistency. It was an assumption known as ω-consistency (also called omega consistency), which we will explain in the next section. Well, in 1936, the logician J. Barkley Rosser came out with another incompleteness proof that did not require Gödel's assumption of ω-consistency, but required only that the system be consistent. We will now explain an essential idea behind Rosser's proof.

The essential idea in question is that to obtain an undecidable sentence, instead of representing P^* (as did Gödel), it suffices to represent some superset of P^* disjoint from R^* (i.e to represent some set A such that P^* is a subset of A, and A has no elements in common with R^*). It also suffices to represent some superset of R^* that is disjoint from P^*, which is in fact what Rosser did.

Theorem R_0 [A forerunner of Rosser's Theorem]. If some superset of R^* disjoint from P^* is representable, or if some superset of P^* disjoint from R^* is representable, then some sentence is undecidable.

Problem 7. (a) Prove Theorem R_0.
(b) Why is Theorem R_0 a strengthening of Theorems G_0 and S_0? That is, why do theorems G_0 and S_0 follow immediately from Theorem R_0?

Separability

Given two disjoint number sets A and B, we say that a formula F *weakly separates* A from B if F represents some superset of A that is disjoint from B. We say that F *strongly separates* A from B if F represents some superset of A and F' represents some superset of B – in other words, if $F(n)$ is provable for every n in A, and $F'(n)$ is provable for every n in B. A is said to be weakly (strongly) separable from B if some formula F weakly (strongly) separates A from B.

Theorem R_0, re-stated, is that if P^* is weakly separable from R^*, or if R^* is weakly separable from P^*, then some sentence is undecidable.

Problem 8. Which, if either, of the following statements are true?

(a) If A is weakly separable from B and the system is consistent, then A is strongly separable from B.
(b) If A is strongly separable from B and the system is consistent, then A is weakly separable from B.

From the answer to the above problem, and from Theorem R_0, we have:

Theorem R_1 [After Rosser]. If R^* is strongly separable from P^*, or if P^* is strongly separable from R^*, and if the system is consistent, then some sentence is undecidable.

What Rosser did was to strongly separate R^* from P^*. How he did this will be explained later on.

Omega Consistency (ω-Consistency)

Imagine that we are all immortal, but that there is a disease which, if caught, puts one in a deep sleep forever. There is also an antidote which, if taken, wakes you up for a limited time, but then you fall back to sleep forever, and no antidote is then effective. The antidote works as follows. If you take it today (the first day), you wake up for two days, and then lapse back to sleep. If taken tomorrow (the second day), you wake up for four days, and then lapse back to sleep. For any positive number n, if you take the antidote n days from now, you remain awake for 2^n days. Now, suppose your loved one has caught the disease, and it is up to you to decide when you should give her the antidote. You of course want to have her awake for as many days as possible. On any day you think, "I shouldn't give her the antidote today. If I only wait until tomorrow and then give her the antidote, I will have her here twice as long!" Thus, on any day it is irrational to give her the antidote on that day, yet it is certainly irrational never to give her the antidote at all!

This situation may well qualify as an example of an omega inconsistency, which roughly speaking is this: Consider a system that proves various facts about the natural numbers. Suppose that the system can prove that there exists a number having a certain property, yet for every particular number n, it can be proved that n doesn't have the property. Such a system would be called *omega inconsistent* (or equivalently, *ω-inconsistent*). [Note that ω is the lower-case Greek letter with the corresponding upper-case letter Ω. As indicated by our use of "ω" here, the Greek letter ω, and Ω as well, are both read "o-me′-ga".]

The situation can be analyzed as follows. Again imagine that we are all immortal, and imagine that there are infinitely many banks in the universe – Bank 1, Bank 2, ..., Bank n, ... You are given a check which says, "Payable at some bank." You go to Bank 1, but they cannot cash it, you then go to Bank 2, Bank 3, and so on, but none of them can cash it. At no time, not even after you have tried a million banks, can you prove that the check is bad, because it is possible that some bank you visit in the future might cash it.

A humorous illustration of ω-inconsistency was given by the late mathematician Paul Halmos. He defines an omega inconsistent mother as one who says to her child, "You can't do this, you can't do that, you can't do ...." When the child asks, "Isn't there *anything* I can do?" the mother replies, "Yes, there is something you can do, but it's not this, not that, not ...."

A precise definition of omega consistency will be given in the next section.

First-Order Systems

We now consider a system based on First-Order Logic. That is, we take some complete axiom system of First-Order Logic, and add additional axioms about the natural numbers. We assume the system to be closed under the rule of modus ponens, that is, if X and $X \supset Y$ are both provable, so is Y. From this it easily follows that for any finite set S of sentences and any sentence X, if X s a first-order consequence of S (i.e. true in all interpretations in which all elements of S are true), then if all elements of S are provable, so is X.

In this system each natural number n has an expression denoted $\overline{n}$ that serves as a *name* of the number n. These names $\overline{0}, \overline{1}, \ldots, \overline{n}, \ldots$, are called *numerals*. For any formula $F(x)$ with x as the only free variable, and any number n, by $F(\overline{n})$ is meant the result of substituting the numeral $\overline{n}$ for all free occurrences of x in $F(x)$. What we have previously written $F(n)$ will now be written $F(\overline{n})$.

If $F(\overline{n})$ is provable, then of course so is $\exists x F(x)$, since the sentence $F(\overline{n}) \supset \exists x F(x)$ is valid. Now, we call the system *omega inconsistent* (or *ω-inconsistent*) if there is a formula $F(x)$ such that $\exists x F(x)$ is provable, and yet all the infinitely many sentences $\sim F(\overline{0}), \sim F(\overline{1}), \ldots, \sim F(\overline{n}), \ldots$, are provable. Now, it is obvious that *in any interpretation in the domain of the natural numbers*, at least one of the infinitely many sentences $\exists x F(x), \sim F(\overline{0}), \sim F(\overline{1}), \ldots, \sim F(\overline{n}), \ldots$, must be false (because if $\exists x F(x)$ is true, then there must be at least one n such that $F(\overline{n})$ is true, in which case $\sim F(\overline{n})$ is false). Nevertheless, one cannot derive a formal contradiction from that set of formulas, i.e. one cannot derive from that set some sentence and its negation. The reason why we cannot get a formal inconsistency from that infinite set is that a proof is only of finite length, hence can use only finitely many of those sentences. Indeed it is known that there are consistent systems that are ω-inconsistent.

A system is called *omega consistent* (or *ω-consistent*) if it is not omega/ω-inconsistent. Thus to say that a system is ω-consistent is to say that for every formula $F(x)$, if $\exists x F(x)$ is provable, then there must be at least one number n such that $F(\overline{n})$ is not refutable.

When talking about ω-consistency, to avoid possible confusion, what we have previously called *consistency* is alternatively called *simple consistency*.

Problem 9. We have already remarked that an ω-inconsistent system can be simply consistent. Is an ω-consistent system necessarily simply consistent?

Representability and Definability

By definition, to say that $F(x)$ *represents* a number set A is to say that for all numbers n, the sentence $F(\overline{n})$ is provable if and only if $n \in A$.

We now say that $F(x)$ *defines* A to mean that for every number n the following two conditions hold:

D_1: If $n \in A$ then $F(\overline{n})$ is provable.
D_2: If $n \notin A$ then $F(\overline{n})$ is refutable.

We say that A is *definable* (in the system) if some formula $F(x)$ defines A.

We say that $F(x)$ *completely represents* A if $F(x)$ represents A and $\sim F(x)$ represents the complement $\widetilde{A}$ of A.

Problem 10. Which, if any, of the following statements are true?

(1) If F completely represents A, then F defines A.
(2) If F defines A, then F completely represents A.
(3) If F defines A and the system is simply consistent, then F completely represents A.

Consider now a formula $F(x, y)$ with x and y as the only free variables, and let x be designated to be the first free variable in the formula $F(x, y)$ and let y be designated the second free variable in the formula. [In practice, the individual variables are arranged in an infinite sequence $v_1, v_2, \ldots, v_n, \ldots$, and when we say that x is designated as the first and y as the second variable in a formula, we mean that for some i and j, $i < j$ and $x = v_i$ and $y = v_j$.] For any numbers n and m, by $F(\overline{n}, \overline{m})$ is meant the result of substituting the numeral $\overline{n}$ for all free occurrence of x and $\overline{m}$ for all free occurrences of y.

We say that a formula $F(x, y)$ *defines* a binary relation $R(x, y)$ if for all numbers n and m, if $R(n, m)$ holds, then $F(\overline{n}, \overline{m})$ is provable, and if $R(n, m)$ doesn't hold, then $F(\overline{n}, \overline{m})$ is refutable. [Similarly we will say that an n-ary relation $R(x_1, \ldots, x_n)$ is said to be *definable* in a system $\mathcal{S}$ if there is a formula $F(x_1, \ldots, x_n)$, said to define R, such that for all numbers $a_1, \ldots, a_n$, the sentence $F(\overline{a_1}, \ldots, \overline{a_n})$ is provable if $R(a_1, \ldots, a_n)$ holds, and is refutable if $R(a_1, \ldots, a_n)$ doesn't hold.]

For any binary relation $R(x, y)$, by the *domain* of R is meant the set of all numbers n such that $R(n, m)$ holds for at least one number m. Also, the *range* of R is defined as the set of all numbers m such that $R(n, m)$ holds for at least one number n.

Enumerability in the System

We shall say that a formula $F(x, y)$ *enumerates* a number set A in the system if for every number n:

(1) If $n \in A$, then $F(\overline{n}, \overline{m})$ is provable for at least one m.
(2) If $n \notin A$, then $F(\overline{n}, \overline{m})$ is refutable for every m.

We say that A is enumerable in the system if some formula $F(x, y)$ *enumerates* A.

A formula $F(x_1, \ldots, x_n)$ is called *numeralwise decidable* if for all numbers $a_1, \ldots, a_n$, the sentence $F(\overline{a_1}, \ldots, \overline{a_n})$ is decidable.

Problem 11. Suppose $F(x, y)$ *enumerates* some set A. Is the formula $F(x, y)$ numeralwise decidable?

Problem 12. Prove that the domain of any definable relation $R(x, y)$ is enumerable in the system. More specifically, show that if the formula $F(x, y)$ defines the relation $R(x, y)$, then $F(x, y)$ enumerates the domain of $R(x, y)$.

The following is a key lemma:

Lemma 2. (a) If A is enumerable in the system $\mathcal{S}$ and $\mathcal{S}$ is omega consistent, then A is representable in $\mathcal{S}$.
(b) More specifically, suppose that $F(x, y)$ enumerates A. Then for any number n:

(1) If $n \in A$, then $\exists y F(\overline{n}, y)$ is provable.
(2) If $\exists y F(\overline{n}, y)$ is provable and if the system is omega consistent then $n \in A$.
(3) Hence, if the system is omega consistent, then the formula $\exists y F(x, y)$ represents A.

Problem 13. Prove Lemma 2.

The Essence of Gödel's Proof

Theorem G_1 [After Gödel]. Suppose the formula $A(x, y)$ enumerates the set P^* and that p is the Gödel number of the formula $\forall y {\sim} A(x, y)$, and that G is the diagonalization, $\forall y {\sim} A(\overline{p}, y)$, of the formula $\forall y {\sim} A(x, y)$. Then:

(a) If the system is simply consistent, then G is not provable.
(b) If the system is omega consistent, then G is also not refutable – hence is undecidable.

Problem 14. Prove Theorem G_1.

It of course follows from Theorem G_1 that if P^* is enumerable in the system and the system is omega consistent, then there is an undecidable sentence. But this weaker fact can be proved far more swiftly using earlier proved results.

Problem 15. How?

Omega Incompleteness

A simply consistent system in which P^* is enumerable by a formula $A(x, y)$ has an even more curious property than incompleteness – namely that there is a formula $H(y)$ such that all the infinitely many sentences $H(\overline{0}), H(\overline{1}), \ldots, H(\overline{n}), \ldots$, are provable, yet the universal sentence $\forall y H(y)$ is not provable! This condition is known as ω-incompleteness (or omega incompleteness).

Such a formula $H(y)$ is $\sim A(\overline{p}, y)$, where p is the Gödel number of $\forall y \sim A(x, y)$. We have seen that $\forall y H(y)$ – which is $\forall y \sim A(\overline{p}, y)$ – is not provable (assuming simple consistency). $\forall y H(y)$ is $F_p(\overline{p})$, and since it is not provable, it is not provable at any stage, hence all the sentences $\sim A(\overline{p}, \overline{0}), \sim A(\overline{p}, \overline{1}), \ldots, \sim A(\overline{p}, \overline{n}), \ldots$, are provable. Thus $H(\overline{0}), H(\overline{1}), \ldots, H(\overline{n}), \ldots$, are all provable, yet $\forall y H(y)$ is not provable.

I might remark that most mathematicians have heard of Gödel's theorem and know of the existence of undecidable sentences, even though few know the proof. But many of them – at least those to whom I have spoken – haven't even heard of the fact that those systems are ω-incomplete, and find this even more surprising!

∃-Incompleteness

Systems which are ω-incomplete have an equally surprising property, which I like to call ∃-incompleteness – namely, that there exists a formula $F(x)$ such that $\exists x F(x)$ is provable, yet none of the sentences $F(\overline{0}), F(\overline{1}), \ldots, F(\overline{n}), \ldots$, are provable.

To see this, note that for any formula $\varphi(x)$ the sentence $\exists x(\varphi(x) \supset \forall y \varphi(y))$ is valid (which the reader can check with a tableau), hence provable. Assuming the system is ω-incomplete, we take for $\varphi(y)$ a formula $H(y)$ such that $\forall y H(y)$ is not provable, yet $H(\overline{n})$ is provable for every n. We now take $F(x)$ to be the formula $H(x) \supset \forall y H(y)$. The sentence $\exists x(H(x) \supset \forall y H(y))$ is valid, hence provable. However, for no n is it the case that $H(\overline{n}) \supset \forall y H(y)$ is provable. This is because $H(\overline{n})$ is provable, so that if $H(\overline{n}) \supset \forall y H(y)$ were provable, then by modus ponens $\forall y H(y)$ would be provable, which it isn't. Thus $\exists x F(x)$ is provable, yet for no n is it the case that $F(\overline{n})$ is provable.

The Rosser Construction

In the systems subject to Gödel's and Rosser's proof, the sets P^* and R^* are enumerable, and also there is a formula, which we write "$x \leq y$" (for easier readability, instead of $\leq x, y$) satisfying the conditions L_1 and L_2 below. [Informally, for any numbers n and m, the sentence $\overline{n} \leq \overline{m}$ expresses the proposition that n is less than or equal to m.]

L_1: For any formula $F(x)$ and number n, if $F(\overline{0}), F(\overline{1}), \ldots, F(\overline{n})$ are all provable, so is $\forall y(y \leq \overline{n} \supset F(y))$.

L_2: For every n, the sentence $\forall y(y \leq \overline{n} \vee \overline{n} \leq y)$ is provable.

Given two formulas $F_1(y)$ and $F_2(y)$, consider the formulas $\forall y(y \leq \overline{n} \supset F_1(y))$ and $\forall y(\overline{n} \leq y \supset F_2(y))$. Those two latter formulas, together with the formula

$\forall y(y \leq \overline{n} \vee \overline{n} \leq y)$, logically imply the formula $\forall y(F_1(y) \vee F_2(y))$, as the reader can verify, using a tableau. Hence from L_2 we have:

L_2': If the formulas $\forall y(y \leq \overline{n} \supset F_1(y))$ and $\forall y(\overline{n} \leq y \supset F_2(y))$ are both provable, so is $\forall y(F_1(y) \vee F_2(y))$.

We will say that a system obeys the *Rosser conditions* if P^* and R^* are both enumerable and conditions L_1 and L_2 both hold.

We aim to prove:

Theorem *R* [After Rosser]. For any system obeying the Rosser conditions, if the system is *simply consistent*, then there is an undecidable sentence of the system.

Separation Lemma for Sets. For any system $\mathcal{S}$ satisfying conditions L_1 and L_2, and any disjoint number sets A and B,

(a) If A and B are enumerable in the system, then B is strongly separable from A in the system.

(b) More specifically, if $F_1(x, y)$ is a formula that enumerates A, and $F_2(x, y)$ is a formula that enumerates B, then B is separated from A by the formula $\forall y[F_1(x, y) \supset \exists z(z \leq y \wedge F_2(x, z))]$.

Problem 16. Prove the Separation Lemma for Sets.

Problem 17. Now prove Theorem R.

We have defined what it means for a formula $F(x, y)$ to *enumerate* a set A in a system $\mathcal{S}$. For some later chapters, we will need the following definitions and results.

Given a numerical relation $R(x_1, \ldots, x_n)$, we will say that a formula $F(x_1, \ldots, x_n, y)$ *enumerates* the relation $R(x_1, \ldots, x_n)$ (in a system $\mathcal{S}$) if for all numbers $a_1, \ldots, a_n$, the relation $R(a_1, \ldots, a_n)$ holds if and only if there is a number b such that $F(\overline{a_1}, \ldots, \overline{a_n}, \overline{b})$ is provable in $\mathcal{S}$.

We will say that a formula $F(x_1, \ldots, x_n)$ *(strongly) separates* a relation $R_1(x_1, \ldots, x_n)$ from the relation $R_2(x_1, \ldots, x_n)$ if for all numbers $a_1, \ldots, a_n$, if $R_1(a_1, \ldots, a_n)$ holds, then $F(\overline{a_1}, \ldots, \overline{a_n})$ is provable in $\mathcal{S}$, and if $R_2(a_1, \ldots, a_n)$ holds, then $F(\overline{a_1}, \ldots, \overline{a_n})$ is refutable in $\mathcal{S}$.

We say that a relation $R_1(x_1, \ldots, x_n)$ is *disjoint* from a relation $R_2(x_1, \ldots, x_n)$ if there exist no numbers $a_1, \ldots, a_n$ such that $R_1(a_1, \ldots, a_n)$ and $R_2(a_1, \ldots, a_n)$ both hold.

Separation Lemma for Relations. For any system $\mathcal{S}$ satisfying conditions L_1 and L_2, and any disjoint relations $R_1(x_1, \ldots, x_n)$ and $R_2(x_1, \ldots, x_n)$, if the formula

$F_1(x_1, \dots x_n, y)$ enumerates $R_1(x_1, \dots, x_n)$ and the formula $F_2(x_1, \dots x_n, y)$ enumerates $R_2(x_1, \dots, x_n)$, then R_2 is strongly separated from R_1 by the formula

$$\forall y[F_1(x_1, \dots, x_n, y) \supset \exists z(z \leq y \wedge F_2(x_1, \dots, x_n, z))].$$

The proof of the above is just like that of the Separation Lemma for Sets (just replace "x" by "$x_1, \dots, x_n$"), as the reader may verify.

Exercise. For any two relations $R_1(x_1, \dots, x_n)$ and $R_2(x_1, \dots, x_n)$, by $R_1 - R_2$ is meant the relation $R_1(x_1, \dots, x_n) \wedge \sim R_2(x_1, \dots, x_n)$. Obviously $R_1 - R_2$ is disjoint from $R_2 - R_1$. Now, suppose that $\mathcal{S}$ obeys conditions L_1 and L_2 and that R_1 and R_2 are n-ary relations enumerated by $F_1(x_1, \dots x_n, y)$ and $F_2(x_1, \dots x_n, y)$ respectively, that are not necessarily disjoint. Prove that $R_2 - R_1$ is strongly separated from $R_1 - R_2$ by the formula

$$\forall y[F_1(x_1, \dots, x_n, y) \supset \exists z(z \leq y \wedge F_2(x_1, \dots, x_n, z))].$$

Note that the Separation Lemma is but a special case of this, since if R_1 is disjoint from R_2, then $R_1 - R_2 = R_1$ and $R_2 - R_1 = R_2$.

Discussion

I would like to make some comparative remarks about the undecidable sentences of Gödel and Rosser. But first, let us consider a formula $F(x, y)$ that enumerates a set A. Thus a number n is in A iff $F(\overline{n}, \overline{m})$ is provable for some number m, and I would like to call such a number m a *witness* that n is in A (with respect to the formula $F(x, y)$, understood). Now, in the Gödel proof, which assumes omega consistency, there is a formula that enumerates the set P^*, and with respect to that formula, the undecidable sentence of Gödel can be thought of as saying, there is no witness that I am provable," or more briefly, "I am not provable."

By contrast, in the Rosser proof, which assumes only simple consistency, there is a formula that strongly separates R^* from P^*, and with respect to that formula, Rosser's undecidable sentence can be thought of as saying, "For any witness that I am provable, there is a lesser number that is a witness that I am refutable."

This sentence is really a very curious one, and I once asked Rosser how he ever came up with such a strange sentence. To my surprise, he replied, "I didn't plan to eliminate the necessity of Gödel's hypothesis of omega consistency. It's just that I experimented with various alternatives to Gödel's sentence, and when I came up with this one, I suddenly realized what I could do with it!"

I find that most interesting, and I hope it will be better known. It reminds me of a rumor about Gödel's proof that I heard, namely that Gödel did not originally set out to prove the systems incomplete. Rather he set out to prove them inconsistent! He thought he could recreate the liar paradox ("this sentence is false") within the system, but after working awhile, he formalized, not truth, but provability, and thus proved incompleteness, rather than inconsistency.

Solutions to the Problems of Chapter 10

1. For any expression X, the sentence $\sim PN(X)$ is true iff the norm of X is not printable. We now take X to be the expression $\sim PN$, and so the sentence $\sim PN(\sim PN)$ is true if and only if the norm of $\sim PN$ is not printable, but the norm of $\sim PN$ is the very sentence $\sim PN(\sim PN)$! Thus the sentence $\sim PN(\sim PN)$ is true if and only if it is not printable, which means that it is either true and not printable, or not true but printable. The latter alternative is ruled out by the given condition that only true sentences are printable. Thus the sentence $\sim PN(\sim PN)$ is true, but the machine cannot print it.
2. Solving the three parts:
   (a) If H designates A, then the diagonalizer of H must designate $A^{\#}$, for suppose H designates A and K is the diagonalizer of H. Then for any number n, the sentence $K(n)$ is true iff $H(n^*)$ is true, which is the case iff $n^* \in A$, which is true iff $n \in A^{\#}$. Thus $K(n)$ is true iff $n \in A^{\#}$, and so K names $A^{\#}$.
   (b) As for the proof of the lemma, suppose A is nameable. Then $A^{\#}$ is nameable, as we have shown. Let k be the Gödel number of a designator that designates $A^{\#}$. Thus for all numbers n, the sentence $H_k(n)$ is true iff $n \in A^{\#}$. In particular, taking k for n, we see that $H_k(k)$ is true iff $k \in A^{\#}$, which is true iff $k^* \in A$. Thus $H_k(k)$ is true iff its Gödel number k^* is in A, which means that $H_k(k)$ is a Gödel sentence for A.
   (c) As in the proof of Theorem T, we let T be the set of true sentences and let T_0 be the set of Gödel numbers of the elements of T. We are to show that T_0 is not nameable.

   We now consider the complement $\widetilde{T_0}$ of T_0 (i.e. the set of all numbers not in T_0). There obviously cannot be a Gödel sentence for $\widetilde{T_0}$ since such a sentence would be true if and only if its Gödel number is not in T_0, which would mean that the sentence would be true iff its Gödel number was not the Gödel number of a true sentence, which is absurd. Thus $\widetilde{T_0}$ has no Gödel sentence, and since every nameable set does have a Gödel sentence, then $\widetilde{T_0}$ is not nameable. Hence T_0 is not nameable (for if some H named T_0, then H' would name $\widetilde{T_0}$).
3. Suppose n is a formula number. Then n^* is the Gödel number of $F_n(n)$. Hence $n^* \in W^*$ iff the Gödel number of $F_n(n)$ is in W_0, which is true iff $F_n(n) \in W$.
4. Suppose the system is consistent and that F is a formula that represents P^*. Let F_b be the negation F' of F. Since F represents P^*, then $F(b)$ is provable iff $b \in P^*$ iff $F_b(b) \in P$ (since b is a formula number). And this is true iff $F_b(b)$ is provable, which is true iff $F'(b)$ is provable (since F' is F_b), which is then true iff $\sim F(b)$ is provable. Thus $F(b)$ is provable if and only if its negation $\sim F(b)$ is provable. This means that either $F(b)$ and $\sim F(b)$ are both provable, or that neither one is provable. The former alternative is ruled out by the given

fact that the system is consistent. Thus the latter alternative must hold: $F(b)$ is neither provable nor refutable, and is thus undecidable.

5. The proof of this is, if anything, rather simpler than that of Theorem G_0. Suppose the system is consistent and that F represents R^*. Let f be the Gödel number of F. Then $F(f)$ is provable iff $f \in R^*$, which is true iff $F_f(f) \in R$ (by Problem 3), which is true iff $F(f) \in R$ (since F and F_f are the same formula), which in turn is true iff $F(f)$ is refutable. Thus $F(f)$ is provable iff $F(f)$ is refutable, which means that either $F(f)$ is both provable and refutable or neither. Again by the assumption of consistency, $F(f)$ cannot be both provable and refutable, hence it is neither, and is thus undecidable.
6. Yes, it is, and here is how to see it. $F(f)$ is refutable iff $F'(f)$ is provable, which is true iff $f \in P^*$ (since F' represent P^*). And this is true only if $F_f(f)$ is provable, which is true iff $F(f)$ is provable (since f is the Gödel number of F, and hence F_f is F). Thus $F(f)$ is refutable iff $F(f)$ is provable, hence again by the assumption of consistency, $F(f)$ is undecidable.
7. Solution of the two parts:
   (a) First we show that if F is a formula that represents some superset of R^* disjoint from P^*, then $F(f)$ is undecidable, where f is the Gödel number of F.

   Let A be the superset of R^* which is disjoint from P^* and is represented by F. Since F represents A, then $F(n)$ is provable iff $n \in A$. Also, $F(f)$ is provable iff $f \in P^*$. Thus $f \in P^*$ iff $f \in A$. Then, since A is disjoint from P^*, f is in neither P^* nor in A. Since f is not in A, it is not in the subset R^* of A, and thus f is in neither P^* nor R^*, which means that $F(f)$ is neither provable nor refutable.

   Next, suppose that F represents some superset A of P^* disjoint from R^*. Then the complement $\widetilde{A}$ of A is a superset of R^* disjoint from P^*, and is represented by F', and so, as we saw, $F'(f')$ is undecidable, where f' is the Gödel number of F'.

   (b) If the hypothesis of Theorem G_0 holds, so does the hypothesis of Theorem R_0. For suppose the hypothesis of Theorem G_0 holds, i.e. P^* is representable and the system is consistent. By the assumption of consistency, P^* must be disjoint from R^* (because if some number n were in both P^* and R^*, then $F_n(n)$ would be both provable and refutable, contrary to the assumption of consistency). Also, P^* is a superset of itself, hence if P^* is representable, then some superset of P^* (namely P^* itself) is disjoint from R^*, hence the hypothesis of Theorem R_0 holds. Thus Theorem G_0 is but a special case of Theorem R_0. A similar proof shows that Theorem S_0 is but a special case of Theorem R_0.
8. It is statement (b) that is true. For suppose F strongly separates A from B. Then F represents some superset A_1 of A, and F' represents some superset B_1 of B.

Suppose also that the system is consistent. Then A_1 must be disjoint from B_1, for if some number n were in both A_1 and B_1, then $F(n)$ would be provable (since $n \in A_1$) and $F'(n)$ would be provable (since $n \in B_1$), contrary to the assumption of consistency. Thus A_1 is disjoint from B_1. Since A_1 is disjoint from B_1 and B is a subset of B_1, if follow that A_1 is disjoint from B. Thus F represents the superset A_1 of A which is disjoint from B, which means that F weakly separates A from B.

9. Of course it is. If a system is not simply consistent, i.e. if some sentence X and $\sim X$ are both provable, then all sentences are provable, because for any sentence Y, the sentence $X \supset (\sim X \supset Y)$ is a tautology, hence provable, and so if X and $\sim X$ are both provable, then by two applications of modus ponens, Y is provable. Thus if a system is simply inconsistent, all sentences are provable, making the system trivially ω-inconsistent. Since simple inconsistency implies ω-inconsistency, then ω-consistency implies simple consistency.
10. (1) is obviously true. (2) is not necessarily true. (3) is necessarily true, for suppose that F defines A and that the system is simply consistent. Then for any number n:
    (a) If $n \in A$, then $F(\overline{n})$ is provable.
    (b) If $n \notin A$, then $F(\overline{n})$ is refutable.

    We are to show that the converses of (a) and (b) both hold.
    (a)′ Suppose $F(\overline{n})$ is provable. If n were not in A, then $F(\overline{n})$ would be refutable, hence the system would be inconsistent. Since we are assuming consistency, then $n \in A$.
    (b)′ Suppose $F(\overline{n})$ is refutable. If n were in A, then $F(\overline{n})$ would be provable, and the system would be inconsistent. Hence, by the assumption consistency, $n \notin A$.
11. Not necessarily.
12. This is really obvious, for suppose $F(x, y)$ defines the relation $R(x, y)$ and A is the domain of R. Then for any number n:
    (a) If $n \in A$, then $R(n, m)$ hold for some m, hence $F(\overline{n}, \overline{m})$ is provable.
    (b) Suppose $n \notin A$. Then for every m, it is not the case that $R(n, m)$ holds, hence $F(\overline{n}, \overline{m})$ is refutable.

    By (a) and (b), $F(x, y)$ enumerates A.
13. Suppose $F(x, y)$ enumerates A.
    (1) Suppose $n \in A$. Then $F(\overline{n}, \overline{m})$ is provable for some m. Hence $\exists y F(\overline{n}, y)$ is provable.
    (2) Suppose $\exists y F(\overline{n}, y)$ is provable. If n were not in A, then $F(\overline{n}, \overline{m})$ would be refutable for every m, and this, together with the provability of $\exists y F(\overline{n}, y)$, would make the system omega inconsistent! Hence if the system is omega consistent, then it cannot be that n is not in A. Thus n must be in A.
    (3) This is immediate from (1) and (2).

14. Let us first recall from First-Order Logic that for any formula $H(y)$ the sentence $\exists y H(y)$ is provably equivalent to $\sim\forall y \sim H(y)$, and thus $\exists y H(y)$ is provable if and only if $\sim\forall y \sim H(y)$ is provable.

    Now, assume the hypothesis of Theorem G_1. The sentence G is $F_p(\overline{p})$, which is $\forall y \sim A(\overline{p}, y)$.

    (a) Suppose $F_p(\overline{p})$ is provable. The $p \in P^*$, and since $A(x, y)$ enumerates P^*, then $A(\overline{p}, \overline{m})$ is provable for some m. Hence $\exists y A(\overline{p}, y)$ is provable, and therefore, so is $\sim\forall y \sim A(\overline{p}, y)$, which is the sentence $\sim F_p(\overline{p})$. Thus, if G (which is $F_p(\overline{p})$) is provable, so is $\sim G$, and the system is then simply inconsistent. Hence if the system is simply consistent, then G is not provable.

    (b) Suppose the system is omega consistent. Then it is also simply consistent (by Problem 9), hence by (a), the sentence $F_p(\overline{p})$ is not provable, and so $p \notin P^*$. Therefore (since $A(x, y)$ enumerates P^*), $A(\overline{p}, \overline{m})$ is refutable for every m, and so by the assumption of omega consistency, the sentence $\exists y A(\overline{p}, y)$ is not provable. Hence $\sim\forall y \sim A(\overline{p}, y)$ is not provable, and thus $\forall y \sim A(\overline{p}, y)$ (which is the sentence G) is not refutable.

15. This is virtually immediate from Lemma 2 and Theorem G_0: Suppose P^* is enumerable and the system is omega consistent. Then by Lemma 2, the set P^* is representable, and since the system is also simply consistent, there is an undecidable sentence by Theorem G_0.

    We might further add that if the system is omega consistent and $A(x, y)$ is a formula that enumerates P^*, then $\exists y A(x, y)$ represents P^* (again by Lemma 2), hence so does the logically equivalent formula $\sim\forall y \sim A(x, y)$. Thus $\forall y \sim A(x, y)$ is a formula whose negation represents P^*. Then, by Problem 6, if p is the Gödel number of the formula $\forall y \sim A(x, y)$, then the sentence $\forall y \sim A(\overline{p}, y)$ is undecidable, and this sentence is the sentence G. Thus the undecidability of G is an easy consequence of Lemma 2 and Problem 6.

16. We are given that $F_1(x, y)$ enumerates A, and $F_2(x, y)$ enumerates B, and that B is disjoint from A. We are to show that B is strongly separated from A by the formula $H(x)$, namely by $\forall y(F_1(x, y) \supset \exists z(z \leq y \wedge F_2(x, z))$.

    (a) Suppose $n \in B$. Then $n \notin A$ (since A is disjoint from B). Since $n \in B$, then $F_2(\overline{n}, \overline{k})$ is provable for some k. Hence the following formula is provable, since it is a logical consequence of $F_2(\overline{n}, \overline{k})$.

    (1) $\quad \forall y[\overline{k} \leq y \supset \exists z(z \leq y \wedge F_2(\overline{n}, z))]$

    Next, since $n \notin A$, then for every number m, the sentence $F_1(\overline{n}, \overline{m})$ is refutable, and so all the sentences $\sim F_1(\overline{n}, \overline{0}), \ldots, \sim F_1(\overline{n}, \overline{k})$ are provable. Hence by condition L_1, the following is provable:

    (2) $\quad \forall y[y \leq \overline{k} \supset \sim F_1(\overline{n}, y))$

By (2) and (1) and condition L_2', we get

$$\forall y[\sim F_1(\overline{n}, y) \vee \exists z(z \leq y \wedge F_2(\overline{n}, z))]$$

As a consequence, we get the logically equivalent sentence

$$\forall y[F_1(\overline{n}, y) \supset \exists z(z \leq y \wedge F_2(\overline{n}, z))],$$

which is the sentence $H(\overline{n})$. Thus if $n \in B$, then $H(\overline{n})$ is provable.

(b) Next we must show that if $n \in A$, then $H(\overline{n})$ is refutable.

Suppose $n \in A$. Then $n \notin B$. Since $n \in A$, then for some k, the sentence $F_1(\overline{n}, \overline{k})$ is provable. Since $n \notin B$, $F_2(\overline{n}, \overline{0}), \ldots, F_2(\overline{n}, \overline{k})$ are all refutable, hence by condition L_1, the sentence $\forall z(z \leq \overline{k} \supset \sim F_2(\overline{n}, z))$ is provable. Since $F_1(\overline{n}, \overline{k})$ is also provable, so is

(1) $\quad F_1(\overline{n}, \overline{k}) \wedge \forall z[z \leq \overline{k} \supset \sim F_2(\overline{n}, z))$

which is logically equivalent to the following:

(2) $\quad \sim[F_1(\overline{n}, \overline{k}) \supset \exists z(z \leq \overline{k} \wedge F_2(\overline{n}, z))]$

[To see this, note that by Propositional Logic, (1) is logically equivalent to

$$\sim[F_1(\overline{n}, \overline{k}) \supset \sim\forall z(z \leq \overline{k} \supset \sim F_2(\overline{n}, z))]$$

which in turn is equivalent to (2), since $\sim\forall z(z \leq \overline{k} \supset \sim F_2(\overline{n}, z))$ is logically equivalent to $\exists z \sim (z \leq \overline{k} \supset \sim F_2(\overline{n}, z))$, which is logically equivalent to $\exists z(z \leq \overline{k} \wedge \sim\sim F_2(\overline{n}, z))$, which is logically equivalent to $\exists z(z \leq \overline{k} \wedge F_2(\overline{n}, z))$.]

Now, (2) logically implies

(3) $\quad \sim\forall y[F_1(\overline{n}, y) \supset \exists z(z \leq y \wedge F_2(\overline{n}, z))]$

[Since for any formula $\varphi(y)$, the sentence $\forall y\varphi(y) \supset \varphi(\overline{n})$ is valid, hence so is $\sim\varphi(\overline{n}) \supset \sim\forall y\varphi(y)$, and, in particular, this is so for when $\varphi(y)$ is the formula $F_1(\overline{n}, y) \supset \exists z(z \leq y \wedge F_2(\overline{n}, z))$].

Now, (3) is the sentence $\sim H(\overline{n})$, and thus $H(\overline{n})$ is refutable for $n \in A$.

17. We are given that P^* and R^* are enumerable in the system. Assuming the system is simply consistent, the sets P^* and R^* are disjoint (why?), and so by the Separation Lemma, the set R^* is strongly separable from P^*. Thus by Theorem R_1, there is an undecidable sentence of the system.

More specifically, suppose $A(x, y)$ is a formula that enumerates P^* and that $B(x, y)$ is a formula that enumerates R^*. Then by the Separation Lemma, the formula $\forall y(A(x, y) \supset \exists z(z \leq y \wedge B(x, z)))$ strongly separates R^* from P^*, and under the assumption of simple consistency, it weakly separates R^* from P^* (as seen in the solution of Problem 8), and hence, by the solution to Problem 7, its diagonalization

$$\forall y(A(\overline{p}, y) \supset \exists z(z \leq y \wedge B(\overline{p}, z))),$$

[where p is the Gödel number of

$$\forall y(A(x, y) \supset \exists z(z \leq y \wedge B(x, z)))],$$

is an undecidable sentence of the system.

11
Elementary Arithmetic

Preliminaries

Elementary Arithmetic – interchangeably called "First-Order Arithmetic" by mathematical logicians – uses the following fifteen symbols:

$$0 \ ' \ (\ ) \ + \ \times \ \sim \ \wedge \ \vee \ \supset \ \forall \ \exists \ v \ = \ \leq$$

Proof systems of Elementary Arithmetic are based on the logical reasoning we have seen formalized in the systems of Propositional and First-Order Logic of recent chapters, but further include formulas created from the symbols above that will enable the expression of arithmetic truths (as well as of falsehoods and conjectures, of course). And such arithmetic proof systems will always have a set of arithmetic axioms attached to the usual axioms and rules of inference of Propositional and First-Order Logic, arithmetic axioms seen as being correct and as complete as their creators could imagine making them.

In this chapter we are not going to introduce any arithmetic axioms. For that you will have to wait until Chapter 13 in which a set of arithmetic axioms will be included in our Elementary/First-Order Arithmetic to form a system which we will call *Peano Arithmetic*, in honor of Giuseppe Peano, who, in the late 19th century, created the first proof system for the natural numbers. But I expect you to be impressed in this chapter by the *depth of the results* that can be obtained by merely defining *truth* for the formulas of First-Order Arithmetic. Clarifying definitions of truth for the formulas of higher-order formal systems was something first accomplished by Alfred Tarski in a famous paper [1933], which included an extremely important result about Elementary Arithmetic, the proof of which is the primary goal of this chapter. The definition of truth presented here is only for interpretations of Elementary Arithmetic employing the domain of the natural numbers, but mathematicians and logicians have devised ways to define and prove seemingly everything we know about the full set of integers, the rational numbers, and even the real numbers (and beyond in many mathematical directions) just based on truths about the natural numbers. Moreover, it was found that very often the proofs could be carried out within a system of First-Order Arithmetic, although sometimes it was more convenient to go to a higher-order proof system, where the mathematician can quantify over predicates and predicates of predicates, etc.

So now, as we first turn to the syntax of Elementary Arithmetic, let's start learning about the most basic part of any formal system for mathematics, its terms and formulas.

The expressions $0, 0', 0'', 0''', \ldots$, are called *Peano numerals*, and are the names used in Elementary and Peano Arithmetic for the natural numbers $0, 1, 2, 3, \ldots$. Thus the name of the natural number n consists of the symbol 0 followed by n accents. The accent symbol $'$ is to be thought of as standing for "successor" (the successor of n is $n+1$). For any (natural) number n we let $\overline{n}$ be an abbreviation for "the Peano numeral that designates n". For instance, if you see $\overline{4}$ in a formula of Elementary Arithmetic, it is an *abbreviation* for the expression $0''''$. Note, though, that the "bar" function $f(n) = \overline{n}$ can also just be seen as a function from the natural numbers into the set of expressions of First-Order Arithmetic (and it is a function *onto* the subset of the expressions we have just called the *Peano numerals*).

Important Digression: We will employ several *representations* of the natural numbers (0, 1, 2, ...) in this chapter, namely:

(1) our usual decimal notation: 0, 1, 2, 3, 4 ...;
(2) the just-defined *Peano numerals*, which is the only one of the representations of the natural numbers in this chapter that is employed *within* our systems of First-Order and Peano Arithmetic;
(3) a notation, but only for the positive integers in this case, that will be extremely helpful in our metatheory work, one which we will call *dyadic notation*, with each individual number in this notation being called a *dyadic numeral*; this notation is similar to binary notation (which we won't use), but instead of using only the digits 0 and 1 as in binary notation, dyadic notation uses (just) 1's and 2's;
(4) a notation derived from the particular Gödel numbering of all the expressions of Elementary Arithmetic that we will use, but which, when applied to the Peano numerals, can also be seen as just another (easily understandable) notation for the natural numbers.

When we use the word "number" alone in this chapter, we will mean natural number. But to be more specific, when we use the words "number" or "natural number" we are thinking of *the natural numbers abstractly*, i.e. independent of any notation or words we may use to communicate about them. Thus the *abstract* number denoted by 5 in decimal notation is the *concept* every human being has in mind when they think of how many fingers most people have on a hand, so that an abstract number is the *meaning* of the number word a person may be thinking, and is thus independent of the language they think in and of any number notation they might happen to use if they wrote the number down. Thus the *number fact* $2+3=5$ has the same meaning, no matter what language it is expressed in verbally, and no matter what number notation is used for writing it down. In Elementary (First-Order) and Peano Arithmetic we would express that fact with the formula $0'' + 0''' = 0'''''$, in which the *notation* used for the natural numbers is that of the Peano numerals (which notation we could obviously use *outside* our logical system as well as inside it, just as people once counted things by putting notches on a stick, although you can see that Peano numerals would be very difficult to understand once the numbers involved became large, which is one reason the world has settled on decimal notation for most purposes; but remember, even now we have to go to so-called "scientific notation" when the numbers we wish to use are very, very large or very, very small, e.g. the distance from Earth to Pluto is 5.906×10^9 kilometers, and the mass of a proton is 1.65×10^{-24} grams).

Perhaps a better way to describe what we mean by the "abstract number 5" would be to call it the "conceptual number 5", since calling 5 "abstract" makes it sound as if it's not real. But the truth is, what we mean by the abstract/conceptual number n is just the number n with the *meaning* you always have mind when you use number words or notations for numbers. The abstract number 5 is the *real* number 5, and the words and notations for it are just communication artifacts. (You had to have mastered the *concept* of 5 before you could learn to correctly attach a word or notation to it as a young child.)

Each one of the notations above that we will use for the natural numbers is individually very easy to understand, but keeping the uses of the different representations of numbers straight in our metatheory work – and sometimes more than one will be used in the same context! – will be part of the hard work involved in this chapter. Sorry for such long philosophical digression, but it may help later.

Getting back to First-Order (Elementary) Arithmetic again now, the symbols $+$ and $\times$ stand, as usual, for plus and times (addition and multiplication). An apostrophe following an expression will signify the operation of adding 1, i.e. the *successor operation*. The symbol $=$ stands for equality, the symbol $\leq$ for "is less than or equal to."

We need infinitely many expressions called *variables*, which, roughly speaking, stand for unspecified natural numbers. However, we wish to stay within a finite alphabet, and so will take as variables the expressions (v), (vv), (vvv), etc. Thus by a *variable* we shall mean a string of v's enclosed in a pair of parentheses. We shall use the letters x, y, z as standing for unspecified variables.

Notice that we have not mentioned here the *parameters* that were used both in our tableau proof systems and our axiomatic proof systems in the chapters on First-Order Logic. In fact, two reasons those parameters were introduced there are because tableau systems for First-Order Logic are more understandable with the use of parameters, and including them in axiomatic systems for First-Order Logic not only does not make axiomatic First-Order Logic less understandable, but, and which was very important for the metatheory carried out in Chapters 8 and 9, makes the comparisons between tableau and axiomatic proof systems we used easier to describe. But we are now going to focus on Axiomatic First-Order Arithmetic, where parameters have traditionally not been used in axiomatic systems, their roles being replaced there by the use of free variables, as well as by the very important mathematical terms when we get to arithmetic systems. Suffice it to say that if we eliminated the mathematical axioms from the proof system for Peano Arithmetic (soon to be presented in Chapter 13) and required all terms there to simply be free variables, it would be easy to show that we would have a parameter-free system identical in proof capabilities to the axiomatic system presented for First-Order Logic earlier, but now using free variables in place of parameters. Thus all the results of Chapters 8 and 9 for First-Order Logic will hold for our systems of Elementary and Peano Arithmetic as well.

Terms of Elementary / First-Order Arithmetic

An expression is called a *term* if its being so is a consequence of the following rules.

(1) Every Peano numeral $\overline{n}$ is a term, and every variable x is a term.
(2) If t_1 and t_2 are terms, so are $(t_1 + t_2)$, $(t_1 \times t_2)$, and t_1'.

By a *constant term* is meant a term in which no variable appears. Note that this means that besides the arithmetic operators for addition, multiplication and successor, the only other strings of symbols occurring in constant terms are those for Peano numerals.

Designation

Each constant term *designates* a unique natural number in accordance with the following rules:

(1) The numeral $\overline{n}$ designates the number n [for example $\overline{5}$, which is *an abbreviation* for the Peano numeral $0'''''$, designates the (abstract) natural number that is written in decimal notation as 5].
(2) If the term t_1 designates n and the term t_2 designates m, then the term $(t_1 + t_2)$ designates n plus m and $(t_1 \times t_2)$ designates n times m and t_1' designates n plus 1.

Formulas

By an *atomic formula* is meant an expression of the form $t_1 = t_2$ or of the form $t_1 \leq t_2$, where t_1 and t_2 are terms. Such a formula is called an *atomic sentence* if t_1 and t_2 are both constant terms. Starting with the atomic formulas, the class of *formulas* is built up in the same way as in Chapter 9, that is, for any formulas F and G, and any variable x, the expressions $\sim F$, $(F \wedge G)$, $(F \vee G)$, $(F \supset G)$, $\forall x F$, $\exists x F$ are all formulas.

The notions of free and bound occurrences of variables in formulas are the same as those of Chapter 9. Also, the notion of substitution of terms for free occurrences of variables is like that of Chapter 9, with terms taking the place of parameters. That is, as in Chapter 9, we let $\varphi(x)$ be a formula possibly containing the variable x, and for any term t, by $\varphi(t)$ is meant the result of substituting the term t for all free occurrences of x in $\varphi(x)$. Of course, if x doesn't occur in the formula $\varphi(x)$, or occurs only as a bound variable, the result of the substitution will just be $\varphi(x)$ once more. We let $v_1, v_2, v_3, \ldots$, be abbreviations of (v), (vv), (vvv), $\ldots$, respectively, and we say that a variable v_n *precedes* the variable v_m if $n < m$. If we let $\varphi(x, y)$ be a formula possibly containing the variables x and y, where x precedes y in our listing of variables, then for any numerals m and n, by $\varphi(\overline{m}, \overline{n})$ we mean the result of substituting $\overline{m}$ for all free occurrences of x and substituting $\overline{n}$ for all free occurrences of y. And similarly for formulas with more than two free variables.

Truth

In defining the concept of truth here, we are considering interpretations only over the domain of the natural numbers.

A sentence (i.e. a formula with no free variables) is called *true* iff its being so is a consequence of the following rules:

(1) An atomic sentence $t_1 = t_2$ (where t_1 and t_2 are constant terms) is true iff t_1 and t_2 designate the same natural number. An atomic sentence $t_1 \leq t_1$ is true iff the number designated by t_1 is less than or equal to that designated by t_2.

(2) As is Propositional Logic, $\sim X$ is true iff X is not true; $X \wedge Y$ is true iff X and Y are both true; $X \vee Y$ is true iff either X or Y is true; $X \supset Y$ is true iff X is not true or Y is true.

(3) $\forall x \varphi(x)$ is true iff $\varphi(\overline{n})$ is true for every n; $\exists x \varphi(x)$ is true iff $\varphi(\overline{n})$ is true for at least one n.

Arithmetic Sets and Relations

A formula $\varphi(x)$ is said to *express* the set of all numbers n such that $\varphi(\overline{n})$ is true. Thus, for any number set A, to say that $\varphi(x)$ expresses A is to say that for all numbers n, the sentence $\varphi(\overline{n})$ is true iff $n \in A$. For a binary relation R, we say that a formula $\varphi(x, y)$ expresses R if for all numbers n and m, the sentence $\varphi(\overline{n}, \overline{m})$ is true iff $R(n, m)$. Similarly with n-ary relations with $n > 2$.

A set of numbers or a relation between numbers is called *arithmetic* if it is expressed by some formula. Note that when "arithmetic" is used in this particular adjectival sense, the accent is on the third syllable: a-rith-*me*'-tic, as opposed to a-*rith*'-me-tic (the latter being the name of the subject studied in elementary school all over the world, which use of the word retains its accent on the second syllable when used as an adjective, as in "my second grade arithmetic class").

You should realize that sets of natural numbers and relations (binary or n–ary, between pairs or n-tuples of natural numbers) that we will be discussing frequently are also abstract things, conceptual things. When we talk about the set of all even numbers, we are thinking of all those conceptual numbers that are evenly divisible by 2. So those sets and relations are *outside* our formal system, but we work very hard to have expressions within the formal system that *express* the meaning of important conceptual sets and relations. You will begin to see how we do that in Problem 1 right now (where the expressions such as x *div* y in the problem are not themselves expressions in Elementary Arithmetic, but are rather abbreviations for the words that follow them in ordinary mathematics-related English).

Problem 1. Show that the following relations among pairs of (natural) numbers, as well as the set of prime numbers, are arithmetic:

(a) x *div* y (x divides y, i.e. x divides y evenly, that is, with no remainder)
(b) x *pdiv* y (x properly divides y, i.e. x divides y evenly and x is neither 1 nor y)
(c) $x < y$ (x is less than y)
(d) *prm* x (x is a prime number)

We shall say that a set or relation is *expressible in terms of plus and times* if it is expressed by a formula in which the symbol $\leq$ does not appear.

Problem 2. Is every arithmetic set or relations expressible just in terms of plus and times?

It was discovered by Kurt Gödel that the exponential relations $x = y^z$ is arithmetic. We shall later give a proof of the related fact that $x = 2^y$ is arithmetic, which will be sufficient for our purposes here, but you will be given a hint in an exercise as how to show Gödel's full result.

Our overall plan is this. We shall soon introduce what is called a *Gödel numbering*, a way of associating a distinct positive integer with each expression of Elementary Arithmetic. And we will prove Tarski's famous result that the set of Gödel numbers of the *true sentences* is not arithmetic. (Can you see that this result is independent of what arithmetical axioms we might want to add to our logical structure for First-Order Arithmetic?)

As already mentioned, in the chapter after next (Chapter 13), we will consider a famous axiom system for arithmetic known as *Peano Arithmetic*, and a very important result that we will obtain there is that the set of Gödel numbers of the sentences *provable* in the system *is* arithmetic, from which it follows that some true sentence is not provable in the system (assuming that only true sentences are provable). We thus obtain a proof of Gödel's Theorem for Peano Arithmetic.

Indeed, we will exhibit a formula that expresses the set of Gödel numbers of the provable sentences, from which we can exhibit a sentence that is true but not provable in the system.

We will also be considering Gödel's original proof, based on the assumption of omega consistency, and Rosser's proof, based on the weaker assumption of simple consistency.

In preparation for all this, we need the following:

Dyadic Numerals

It is the *dyadic numerals*, mentioned above, that we will use to do the numbering of formulas that is so critical to Gödel's Incompleteness proof. We will now explain what these numerals are. Remember, though, that the dyadic numerals are just another *notation* for numbers, in their case a notation only for the positive integers $1, 2, 3, \ldots$.

Some of you are familiar with *binary notation*, in which every natural number is represented by a string of the digits 1 and 0. For the digits $d_0, d_1, \ldots, d_n$ (all of these digits being a 1 or a 0), the binary number $d_n d_{n-1} \cdots d_1 d_0$ designates the number

$$d_n \times 2^n + d_{n-1} \times 2^{n-1} + \cdots + d_1 \times 2^1 + d_0 \times 2^0.$$

For example, the binary numeral 101101 designates

$$1 \times 2^5 + 0 \times 2^4 + 1 \times 2^3 + 1 \times 2^2 + 0 \times 2^1 + 1 \times 2^0 = 2^5 + 2^3 + 2^2 + 1.$$

This is $32+8+4+1 = 45$ in our usual base 10 notation.

In my book *Theory of Formal Systems* [Smullyan, 1961] I introduced a variant of binary notation, which I called *dyadic notation*, and which has advantages over binary notation for certain purposes. Just as any natural number can be uniquely expressed in binary notation by a string of 1's and 0's, so can any *positive* whole number be uniquely expressed as a string of 1's and 2's in dyadic notation. We call the digits 1 and 2 the *dyadic digits*. Any string $d_n d_{n-1} \cdots d_1 d_0$ of dyadic digits designates the number calculated in exactly the same way as in binary notation (but now with the dyadic digits 1 and 2 rather than the binary digits 0 and 1):

$$d_n \times 2^n + d_{n-1} \times 2^{n-1} + \cdots + d_1 \times 2^1 + d_0 \times 2^0.$$

For example the dyadic numeral 1211 designates

$$1 \times 2^3 + 2 \times 2^2 + 1 \times 2^1 + 1 = 2^3 + 2 \times 2^2 + 2 + 1 = 8 + 8 + 2 + 1 = 19$$

Here are the first 16 positive integers written in dyadic notation:

1	2	3	4	5	6	7	8	9	10	11	12	13	14	15	16
1	2	11	12	21	22	111	112	121	122	211	212	221	222	1111	1112

In dyadic notation, any power of 2 consists of either 2 alone, or a 2 preceded by a string of 1's. A string of 1's of length n denotes 2^{n-1} (in both binary and dyadic notation). Can you see that the dyadic notation for the base 10 number 45 is 12221? We saw above that the binary notation for 45 is 101101. (Such lovely palindromes for 45 in both cases!) The dyadic notation for a number is usually shorter by one digit than the binary notation of the number, except for numbers of the form 2^{n-1}, where both notations are the same, but that is not an advantage of this system of notation that is of interest to us, although in fact certain *length* properties of dyadic numerals will be of interest in a very important proof.

Concatenation of Dyadic Numerals

For any two positive integers x and y, by $x*y$ we shall mean the positive integer designated by the dyadic numeral for x directly followed by the dyadic numeral for y. For example, $5*13 = 53$, because the notations for the numbers 5, 13 and 53 expressed here in decimal notation are, in dyadic notation, 21, 221 and 21221, respectively. Whenever we place one string of symbols X directly after another string of symbols Y to produce a third string of symbols XY, we say we have *concatenated* X and Y, and we will call the symbol $*$ the symbol for *the concatenation operation*. The *concatenation relation* between three numbers will be written $x*y = z$. Thus the *concatenation relation* between positive integers, which will be very important to us now, can be expressed in words as follows: If x and y are positive integers, $x*y$ will mean the result of putting x and y in dyadic notation (i.e. from whatever notation we received them in, which might already have been dyadic notation, but could have been decimal notation, as in the $5*13$ above), and then putting the dyadic notation for y directly after the dyadic notation for x, to form the dyadic notation for a new number (in dyadic notation, which, of course, because of our life-long habits of using decimal notation for our integers, we might want to convert back to decimal notation). In any case, the equation $x*y = z$ is a relation between three conceptual positive integers, independent of notation, but it is clearly best, whenever we encounter numbers being operated on by the concatenation function $*$, to think of those numbers in dyadic notation.

We wish to show that the relation $x*y = z$ (as a relation between three positive integers x, y and z) is arithmetic. In fact we need to do something even stronger, which is necessary for the next chapter.

$\sum_0$ *Relations*

We define a $\sum_0$ formula and a $\sum_0$ relation as follows: By an *atomic* $\sum_0$ formula we mean any formula of one of the forms $c_1 = c_2$ or $c_1 \leq c_2$, where c_1 and c_2 are both terms. We then define the class of $\sum_0$ formulas of Elementary Arithmetic by the following inductive scheme:

1. Every atomic $\sum_0$ formula is a $\sum_0$ formula.
2. For any $\sum_0$ formulas F and G, the formulas $\sim F$, $F \wedge G$, $F \vee G$, and $F \supset G$ are $\sum_0$ formulas.
3. For any $\sum_0$ formula F, and any variable x, and any c which is either a Peano numeral or a variable distinct from x, the expressions $\forall x(x \leq c \supset F)$ and $\exists x(x \leq c \supset F)$ are $\sum_0$ formulas.

No formula is a $\sum_0$ formula unless its being so is a consequence of (1), (2) and (3) above.

We shall frequently abbreviate $\forall x(x \leq c \supset F)$ by $(\forall x \leq c)F$ and $\exists x(x \leq c \supset F)$ by $(\exists x \leq c)F$. We read $(\forall x \leq c)F$ as "For all x less than or equal to c, F holds." Similarly, we read $(\exists x \leq c)F$ as "For some x less than or equal to c, F holds."

The quantifiers $(\forall x \leq c)$ and $(\exists x \leq c)$ are called *bounded quantifiers*. Thus a $\sum_0$ formula is a formula of Elementary Arithmetic in which all quantifiers are bounded. A set or relation is called a $\sum_0$ set or relation (or just "$\sum_0$") if it is expressed by a $\sum_0$ formula. $\sum_0$ relations and sets are also called *constructive arithmetic* relations and sets. Clearly all $\sum_0$ relations and sets are *arithmetic*, because the $\sum_0$ formulas that express them are just a subset of all the formulas of Elementary Arithmetic that can be used to express sets and relations.

Discussion

Given any $\sum_0$ sentence (a $\sum_0$ formula with no free variables), we can effectively decide whether it is true or false. This is obvious for atomic $\sum_0$ sentences. Also, given any two sentences X and Y, if we know how to determine the truth values of X and Y, we can obviously determine the truth or falsity of each of $\sim X$, $X \wedge Y$, $X \vee Y$, and $X \supset Y$. Now, let us consider the quantifiers. Suppose we have a formula $F(x)$ with x as the only free variable. Suppose also that for each number n, we can determine whether $F(\overline{n})$ is true or false. Can we then determine the truth value of $\exists x F(x)$? Not necessarily. If it is true, we can sooner or later know it, by systematically testing the sentences $F(\overline{0})$, $F(\overline{1})$ $F(\overline{2}), \ldots, F(\overline{n}), \ldots$, and if $\exists x F(x)$ is true, then $F(\overline{n})$ will be true for some n. But if $\exists x F(x)$ is false, our search will be endless: at no point will we come across a true sentence $F(\overline{n})$, and at no point can we know that we won't come across such a true $F(\overline{n})$ in the future. Thus if $\exists x F(x)$ is true, we will sooner or later know it, but if it is not true, we may never know it. As for the universal quantifier $\forall x F(x)$, if it is false, we will sooner or later know it (by successively testing $F(\overline{0})$, $F(\overline{1})$ $F(\overline{2}), \ldots, F(\overline{n}), \ldots$,), but if it is true, we may never know it.

The situation is very different with bounded quantifiers. Consider a sentence $(\exists x \leq \overline{n})F(x)$ for some number n. We only have to test the finitely many sentences $F(\overline{0})$, $F(\overline{1})$ $F(\overline{2}), \ldots, F(\overline{n})$, to determine the truth value of $(\exists x \leq \overline{n})F(x)$, and similarly with $(\forall x \leq \overline{n})F(x)$. Thus for any $\sum_0$ sentence we can effectively determine whether it is true or false.

We now wish to show that the concatenation relation $x*y = z$ is not only arithmetic, but is in fact $\sum_0$.

Let us first note that the relation x *div* y (x divides y evenly) is not only arithmetic, but is even $\sum_0$, for it can be expressed as follows: $(\exists z \leq y)(x \times z = y)$. Also, *prm* x is $\sum_0$ because another way to express it is $(\forall y \leq x) \sim y$ *pdiv* x. With these comments, you should be able to see that all the relations Problem 1 are not only arithmetic but also have the stronger property of being $\sum_0$ (constructive arithmetic).

Note: Actually, as a $\sum_0$ formula in Elementary Arithmetic, *prm* x is $(\forall y \leq x)(\sim(\exists z \leq y)(x \times z = y))$. But we cannot continually write out the full Elementary Arithmetic formulas because they would get too long, and too difficult to understand, so we will assume the reader can substitute the formulas we have shown to have this or that property into formulas containing an abbreviation for the relation they define.

Now, what is the relation $x*y = z$ in terms of plus and times? Not so easy, but here goes! First of all, for any positive integer x, we let $L(x)$ be the *length* of x when x is written in dyadic notation. For instance, the number 19 in dyadic notation is 1211 and the length of the dyadic numeral 1211 is 4, so $L(19) = 4$.

Now, it is not difficult to show that $x*y = x \times 2^{L(y)} + y$: What this notation means is that if x has the dyadic notation $d_j d_{j-1} \cdots d_1 d_0$ and y has the dyadic notation $e_k e_{k-1} \cdots e_1 e_0$, we can find the conceptual number the concatenation of x and y designates by noticing that $L(y)$ in our example is $k+1$, so we first multiply x by 2^{k+1}, which adds $k+1$ to the exponent on each of the powers of 2 in the dyadic expansion of x. And then we add the result to y (looking at its dyadic expansion, in order to see the dyadic expansion of the number that results):

$$d_j \times 2^{j+k+1} + d_{j-1} \times 2^{j-1+k+1} + \cdots + d_1 \times 2^{1+k+1} + d_0 \times 2^{0+k+1}$$
$$+ e_k \times 2^k + e_{k-1} \times 2^{k-1} + \cdots + d_1 \times 2^1 + d_0.$$

We can see that when we write out this result as a dyadic numeral, it is exactly the dyadic expansion of x followed directly by the dyadic expansion of y. Thus $x*y = x \times 2^{L(y)} + y$.

Our job now (on the way to proving that $x*y = z$ is a $\sum_0$ relation, i.e. expressible in First-Order Arithmetic by a $\sum_0$ formula) is to first prove that the relation $x = 2^{L(y)}$ is a $\sum_0$ relation between x and y. [In *many* words, this relation $x = 2^{L(y)}$ holds for any two arbitrary (conceptual) natural numbers x and y iff y is a *positive* integer, and if we denote by $L(y)$ the length of y when y is written out in dyadic notation, then the (conceptual) number x is equal to the (conceptual) number $2^{L(y)}$.]

To prove that $x = 2^{L(y)}$ is $\sum_0$, we first note that for any number r, the smallest number in dyadic notation of length r consists of a string of 1's of length r, which has the value $2^r - 1$. The largest number of length r consists of a string of 2's of length r, which is two times the smallest number of length r, and hence has the value $2 \times (2^r - 1)$. Thus a dyadic numeral y has length r if and only if the following condition holds:

$$(*)\ 2^r - 1 \leq y \leq 2 \times (2^r - 1) \quad \text{(i.e. } y \text{ lies between } 2^r - 1 \text{ and } 2 \times (2^r - 1)\text{)}.$$

From the condition $(*)$ and the fact that $L(y)$ is meaningless unless y is a positive integer, it follows that $x = 2^{L(y)}$ if and only if the following three conditions hold (note that $x \neq 0$ is forced by the satisfaction of condition C_2);

C_1: $x - 1 \leq y \leq 2 \times (x - 1)$.
C_2: x is a power of 2.
C_3: $y \neq 0$.

Problem 3. Using the condition $(*)$, prove that $y > 0$ and $x = 2^{L(y)}$ hold if and only if conditions C_1, C_2 and C_3 all hold.

Note that all the arithmetical reasoning we have just been doing has been *outside* the system of Elementary Arithmetic. We have been doing this mathematical reasoning *about* our logical system in every-day mathematical English. And this reasoning outside the system of Elementary Arithmetic has been about how we can express mathematical truths *inside* the system. When we are doing this, we are doing what we call *metatheory*. Reasoning *about* formal proof systems is what mathematical logic is all about, even though most such reasoning should, with a great deal of work, be able to be formalized in some logical system. But the reasoning in question usually becomes more difficult to understand when formalized. That's not really a problem because the whole point of thinking *about* formalization is to make sure it is correctly catching our ordinary mathematical reasoning, without which it wouldn't have any value. As with conceptual numbers, our conceptual reasoning is really the important thing. Mathematical logic just addresses the important question of what lies at the basis of our conceptual reasoning (i.e. what are our implicit axioms and rules of inference), and formalizing our reasoning makes this clearer. Doing so has also raised many interesting questions, a good number of the important ones of which are being communicated to you in this book. But in doing mathematical logic, one has to keep clear about which proofs (or formulas etc.) are *within* the formal system being discussed (e.g. all those proofs worked out as solutions to problems in Chapter 7 were within the axiomatic system for First-Order Logic) and which proofs (and formulas etc.) are not in a formal system. Realize that the whole definition of truth and the concept of *expressing* a mathematical relation or set within the system of Elementary Arithmetic are themselves in the metatheory, i.e. outside the logical system we are looking at for arithmetic. People often say that the formulas of formal systems have no meaning. But you already have enough experience in metatheory to see that metatheory (whether for Propositional Logic, pure First-Order Logic, or Elementary Arithmetic, our topic now), to be meaningful, must reason about interpretations, the place where formulas of a formal system take on meaning.

Now, it is easy to see that condition C_1 is $\sum_0$.

Problem 4. Why?

But what about C_2? How can we show that the property "x is a power of 2" is $\sum_0$ (constructive arithmetic), or even arithmetic, since we don't yet have the exponential relation $x^y = z$ available? Well, here we use a very clever idea due to the logician John Myhill: 2 is a prime number, and for any prime number p, a number x is a power of p if and only if every divisor of x other than 1 is itself divisible by p!

Problem 5. More relations to be shown to be $\sum_0$:

(a) $Pow_2(x)$ (x is a power of 2).

(b) The relation $0 \neq x \wedge 0 \neq y \wedge x{*}y = z$.

We note that the function $x{*}y$ is *associative*, i.e. that $(x{*}y){*}z = x{*}(y{*}z)$, and hence that parentheses don't matter; we can write either of those expressions as $x{*}y{*}z$. [This, by the way, is not true for the concatenation of numbers written in *binary notation*. For example $(1{*}0){*}1$ is not the same as $1{*}(0{*}1)$ because $0{*}1 = 1$, so that $1{*}(0{*}1)$ is 11, not $(1{*}0){*}1$. This is one technical advantage of dyadic notation over binary notation.]

Problem 6.

(a) Show that the relation $(0 \neq x_1 \wedge 0 \neq x_2 \wedge 0 \neq x_3 \wedge x_1{*}x_2{*}x_3 = y)$ is $\sum_0$ (constructive arithmetic).

(b) Using mathematical induction, show that for any $n \geq 2$, the relation $(0 \neq x_1 \wedge \cdots \wedge 0 \neq x_n \wedge x_1{*}x_2{*}\cdots{*}x_n = y$ is $\sum_0$ (constructive arithmetic).

Dyadic Gödel Numberings

Until further notice, we shall abbreviate $x{*}y$ by xy (not to be confused with x times y, which will be written $x \times y$ in both our written metalanguage and in the formulas of First-Order Arithmetic). Thus xy is the number designated by the concatenation of the two dyadic numerals for x and y.

Dyadic notation provides a technically very convenient scheme of Gödel numbering: For any positive number n, we let g_n be the positive integer which in dyadic notation consists of 1 followed by n occurrences of 2. Thus $g_1 = 12$; $g_2 = 122$, $g_3 = 1222$, etc. Now, to each of the fifteen symbols $0 \; ' \; (\;) \; + \; \times \; \sim \; \wedge \; \vee \; \supset \; \forall \; \exists \; v \; = \; \leq$ we assign the Gödel numbers $g_1, g_2, \ldots, g_{15}$ respectively. For a complex expression formed from these symbols, we take its Gödel number to be the result of replacing each symbol by its Gödel number. For example, the (meaningless) expression "$(+\wedge$", (which is the 3rd symbol, followed by the 5th symbol, followed by the 8th symbol), has the Gödel number $g_3g_5g_8$, which is 122212222122222222. You can easily see that this method of assigning a Gödel number to each expression of Elementary Arithmetic is done in such a way that we can easily get back to any expression from its Gödel number.

As noted earlier, this method of Gödel numbering has the technical advantage that for any expressions X and Y, with respective Gödel numbers x and y, the Gödel number of XY is xy. In mathematical parlance, "Dyadic Gödel numbering is an *isomorphism* with respect to concatenation." For any n, we denote the Gödel number for the expression $\overline{n}$ by $g(\overline{n})$.

We are aiming to show that the set of dyadic Gödel numbers of the *true* sentences is not arithmetic, i.e. that there is no formula $F(x)$ such that for all numbers n, the sentence $F(\overline{n})$ is true iff n is the Gödel number of a true sentence (this is Tarski's Theorem). In the next chapter, we will consider a famous axiom system for arithmetic and show that the set of *provable* sentences *is* arithmetic, from which we will conclude that there is a true sentence that is not provable in the system (assuming that the system is *correct*, in the sense that only true sentences are provable in the system). We will thus obtain one proof of Gödel's incompleteness theorem for the axiom system.

We have one particular hurdle to be overcome in the proof of Tarski's Theorem [1933], and the Gödel numbering of all the expressions of Elementary Arithmetic will be central to it. But before delving into the most difficult part of the proof, let's pause and notice that the Gödel numbering for all expressions just defined happened to give us a new notation for the positive integers, which we could call the *Gödel notation*, and the numbers expressed in this notation the *Gödel numerals*. So, for any expression X, let's denote by $g(X)$ be the Gödel number of X. Now, for any natural number n, the Peano numeral $\overline{n}$ that designates n, like any expression, has a Gödel number $g(\overline{n})$. We note that $g(\overline{0}) = g(0) = 12$; $g(\overline{1}) = g(0') = 12122$; $g(\overline{2}) = g(0'') = 12122122; \ldots, g(\overline{n+1}) = g(\overline{n}) * 122$. Can you see why?

Note that although the Gödel notation is just a part of the Gödel numbering of all expressions, it can be seen independently as just another notation for the positive integers. Here's a table that sums up all our notations for a few numbers.

Base 10 Notation	Dyadic Notation	Peano Notation	Gödel Notation
0	none	0	12
1	1	$0'$	12122
2	2	$0''$	12122122
3	11	$0'''$	12122122122
4	12	$0''''$	12122122122122

However, it is important to point out now that the Gödel number attached to a particular string of characters in Elementary Arithmetic, be that string a Peano Numeral or a formula, will be interpreted as the (abstract) number designated by the Gödel number when read as a dyadic numeral, even when the Gödel number for a Peano Numeral is in question, Thus, in *decimal* notation, the Gödel numbers for the Peano numerals 0, $0'$, $0''$, $0'''$, and $0''''$ are $4, 42, 346$. 3,802, and 22,234, the abstract numbers (expressed here in decimal notation) designated by the Gödel Numbers (taken as dyadic numerals) 12, 12122, 12122122, 12122122122, 12122122122122, respectively.

The most difficult part of the proof of Tarski's Theorem is showing that the relation $g(\overline{n}) = m$ (considered as a binary relation between the two natural numbers n and m) is arithmetic There are several ways this can be done. One well-known way uses a result from number theory known as the Chinese Remainder Theorem. We shall go a different way, using a modification of results due to Quine. We need a good number of preliminaries:

Consider two positive integers x and y. We shall say that x *begins* y if $x = y$ or if the dyadic notation for x is an initial segment of the dyadic notation for y, i.e. $xz = y$ for some z in dyadic notation. We say that x *ends* y if $x = y$ or $zx = y$ for some z. We say that x is *part* of y if x begins y or x ends y or $z_1xz_2 = y$ for some z_1 and z_2. (Note that since all the relations we have just expressed are based on the concatenation relation, and since the requirement for x and y being positive is built into the formula expressing the concatenation relation in Elementary Arithmetic, we don't have to repeat the requirement that the numbers involved be positive integers if we use the concatenation operation in the creation of a $\sum_0$ formula).

Problem 7. Show the following to be $\sum_0$ (for $x \neq 0$ and $y \neq 0$):

(a) xBy (the dyadic notation for x begins the dyadic notation for y)
(b) xEy (the dyadic notation for x ends the dyadic notation for y)
(c) xPy (the dyadic notation for x is part of the dyadic notation for y)

Now consider a finite sequence S of n ordered pairs of *positive integers*: $(a_1, b_1), \ldots, (a_n, b_n)$. We are now going to construct a certain notation for a dyadic numeral z, which will *encode* the whole set of $2n$ numbers, indeed encode them so that we can easily extract all of them as ordered pairs by *decoding* z. We follow a method developed by Quine to do this. First we consider all the positive integers in the sequence and we let t be the shortest string of 1's longer than every string of 1's in the dyadic notation of any of the numbers a_i, b_i. (Thus if none of the dyadic numerals for the positive numbers in the ordered pairs of the sequence have any 1's in them, the shortest possible t will be the string consisting of exactly one 1.) We let $f = 2{*}t{*}2$ (or more simply $2t2$) and we call an f formed this way in relation to the sequence $(a_1, b_1), \ldots, (a_n, b_n)$ the *frame* for the sequence.

Then we take the dyadic numeral notations for the ordered pairs in that sequence and the frame f that we have constructed, and form the following dyadic numeral by concatenation:

$$(^*) \qquad ffa_1fb_1ff --- ffa_nfb_nff.$$

And we call the *positive integer designated by this dyadic numeral z*, and we call such a number z a *code number for the finite sequence of ordered pairs* $(a_1, b_1), \ldots, (a_n, b_n)$.

Note: The expression

$$f{*}f{*}a_1{*}f{*}b_1{*}f{*}f{*} --- {*}f{*}f{*}a_n{*}f{*}b_n{*}f{*}f$$

is the concatenation of the dyadic numerals for the various conceptual positive integers f, a_1, etc. It doesn't matter that f was introduced to us in its dyadic notation from the beginning, and we were probably imagining the a_i's and b_i's in their decimal notation (as we usually do.) Once they all end up in a concatenation, we just have to think of their dyadic notations being used to form the final dyadic numeral.

Now the code number we just defined from our sequence $(a_1, b_1), \ldots, (a_n, b_n)$ is the nicest, most efficient code number we could use to encode our sequence, because it is very easy to extract the sequence from this code number, and everything in the code number is relevant to the encoding. But to express precisely this construction in a formula of Elementary Arithmetic is quite messy, so we would like to use the basic idea embedded in this construction above to allow additional code numbers for the same sequence, although still ones from which we can extract an encoded sequence. (We, being wise, though, if *we* wanted to encode a sequence in a code number, would use precisely the above encoding.) Here's how we will define a code number for a finite sequence of ordered pairs. Let's start by redefine our word *frame* to mean any string of 1's preceded and followed by a 2. Then the critical things that enable the dyadic notation for a positive integer z to be an encoding of a finite sequence of ordered pairs of positive integer $(a_1, b_1), \ldots, (a_n, b_n)$ are the following.

- The longest frame f in z contains a string of 1's longer than any string of 1's in any of the numbers in the ordered pairs $(a_1, b_1), \ldots, (a_n, b_n)$.
- Most importantly, every ordered pair (a, b) in the set is found within z in a string of the form $ffafbff$, where f occurs in neither a nor b, and with f being the longest frame occurring in z.
- The pairs $(a_1, b_1), \ldots, (a_n, b_n)$ are found encoded in the sequence as we move from left to right in the dyadic notation for z.

These conditions guarantee that any positive integer z will be a *code number* for a unique finite sequence of ordered pairs of positive numbers if z satisfies the following two conditions:

(1) z contains at least one frame.
(2) If f is the longest frame in z, then there is at least one string in z of the form $ffafbff$ where f is not a part of a or b.

And when z is a code number satisfying these two conditions, then z is said to encode the sequence of all the ordered pairs of positive integers (a, b) such that $ffafbff$ is part of z (and the order in the sequence is obtained by the order in which this kind of string occurs within z, as one moves from left to right within its dyadic notation).

Note that this "weak" definition of code number allows a code number z for a finite sequence of ordered pairs $(a_1, b_1), \ldots, (a_n, b_n)$ to start or end with a longer string of 1's than are in f, and there can also be strings inside z that are like "junk DNA" – they have nothing to do with the sequence that is encoded in z. For instance any string like $ffafbfcfdff$, with f not being part of any of a, b, c, d would be such an irrelevant part. The *relevant* parts are all those of the form $ffafbff$ where f is not a part of a or b.

Exercise. Prove that if z satisfies the conditions (1) and (2) above there is one and only one sequence of ordered pairs of positive integers encoded in z.

Now let's take a look at the important Lemma we have to prove:

Lemma K_1 [After Quine]. There is a $\sum_0$ relation $K_1(x, y, z)$ having the following two properties:

1. z is a code number for a finite sequence $(a_1, b_1), \ldots, (a_n, b_n)$ of ordered pairs of positive integers, and the relation $K_1(x, y, z)$ holds if and only if (x, y) is one of the pairs $(a_1, b_1), \ldots, (a_n, b_n)$.
2. For any three numbers x, y and z, if $K_1(x, y, z)$ holds, then $x \leq z$ and $y \leq z$.

The subscripts of 1 in the above lemma are to remind us that Lemma K_1 holds for sequences of ordered pairs of *positive* integers (the positive integers start with the integer 1). To prove Lemma K_1 we will need to know that the following relations are $\sum_0$ (assume "the dyadic notation for the positive integer" in front of every variable mentioned):

$ones(x)$ (x is a string of 1's)

iff $\sim 2\,P\,x$

$x\,fr\,z$ (x is a frame of z)

iff $(\exists y \leq z)\,(ones(y) \wedge x = 2y2 \wedge x\,P\,z)$

$lf(x, y, z)$ (x and y are frames of z and x is longer than y)

iff $(\exists v \leq z)(\exists w \leq z)(x\,fr\,z \wedge y\,fr\,z \wedge$

$y = 2v2 \wedge ones(w) \wedge x = 2vw2)$

$x \; max \; z$ (x is the longest frame occurring in z)

iff $x \, fr \, z \wedge (\forall u \leq z)((u \, fr \, z \wedge u \neq x) \supset lf(x, u, z))$

We are finally ready to construct our $\sum_0$ relation for $K_1(x, y, z)$! Here it is:

$$(\exists f \leq z)(f \; max \; z \wedge ffxfyff \; P \; z \wedge \sim f \; P \; x \wedge \sim f \; P \; y)$$

Since we wish to apply Quine's Lemma in a way that involves formulas of Elementary Arithmetic, whose domain of interpretation is always the *natural* numbers, we will actually require a version of Lemma K_1 that speaks about the encoding of finite sequences of ordered pairs of *natural numbers* (which of course start with the number 0). Here is the result we need, which follows easily from Lemma K_1:

Lemma K_0. (Quine's lemma for the natural numbers). There exists a $\sum_0$ relation $K_0(x, y, z)$ such that for any finite sequence S of ordered pairs of *natural numbers*, there is a number z such that for all numbers x and y, the relation $K_0(x, y, z)$ holds if and only if (x, y) is one of the ordered pairs in S.

Problem 8. Now that Lemma K_1 (for sequences of ordered pars of positive integers) has been proved, prove Lemma K_0 for sequences of ordered pars of natural numbers.

$\sum_1$ *Relations*

A relation is called $\sum_1$ if it is of the form $\exists z R(x_1, \ldots, x_n, z)$, where the relation $R(x_1, \ldots, x_n, z)$ is $\sum_0$. Every $\sum_1$ is of course arithmetic.

Now that we have Lemma K_0, it is easy to prove that the relation $g(\overline{x}) = y$ (as a relation between the natural numbers x and y) is $\sum_1$.

Problem 9. Prove that the relation $g(\overline{x}) = y$ is $\sum_1$. [Hint: Consider the sequence of ordered pairs of natural numbers $(0, g(\overline{0})), (1, g(\overline{1})), \ldots, (n, g(\overline{n}))$, and remember that $g(\overline{0}) = 12$, and that for each positive number $i \leq n$, $g(\overline{i+1}) = g(\overline{i}) * 122$.]

Exercise. Prove that the relation $x^y = z$ is $\sum_1$. [Hint: Consider the sequence $(0, a_0), (1, a_1), \ldots, (n, a_n)$, where $a_0 = 1$, $a_n = x^n$ [$a_n = z$ in the equation $x^y = z$] and for each $i < n$, $a_{i+1} = a_i \times x$. Then use Lemma K_0.]

We shall subsequently need to know a key fact about $\sum_1$ relations: Consider a $\sum_0$ relation $R(x, y, z_1, \ldots, z_n)$, and now consider $\exists x \exists y R(x, y, z_1, \ldots, z_n)$, (as a relation between the variables $z_1, \ldots, z_n$). Since this relation involves the two unbounded existential quantifiers $\exists x$ and $\exists y$, it does not appear on the surface to be $\sum_1$, but in fact it is $\sum_1$!

Problem 10. Prove this, and hence that also if a relation $R(x, z_1, \ldots, z_n)$ is $\sum_1$, so is the relation $\exists x R(x, z_1, \ldots, z_n)$.

Tarski's Theorem

Now that we have proved that the relation $g(\overline{x}) = y$ is arithmetic, we are ready to prove Tarski's Theorem for elementary arithmetic.

We shall take one particular variable, say v_1, and consider a formula $F(v_1)$ with v_1 as its only free variable. We have defined $F(\overline{n})$ to be the result of substituting the Peano numeral $\overline{n}$ for all free occurrences of v_1 in the formula $F(v_1)$. Is the Gödel number of $F(\overline{n})$ expressible as an arithmetic function of the Gödel number of $F(v_1)$ and the number n? That is, is there an arithmetic relation $R(x, y, z)$ which holds iff x is the Gödel number of a formula $F(v_1)$ and the Gödel number of $F(\overline{y})$ is z? Yes, there is such an arithmetic relation, but to prove this is quite elaborate, as it involves arithmetizing the process of substitution. Fortunately we can avoid this by a slight modification of a clever idea due to Alfred Tarski [1953]. Informally, the idea is this: To say of a given property P that it holds for a given number n is equivalent to saying that there exists a number x such that $x = n$ and P holds for x. Formally, the sentence $F(\overline{n})$ is equivalent to the sentence $\exists v_1(v_1 = \overline{n} \wedge F(v_1))$. [It is also equivalent to $\forall v_1(v_1 = \overline{n} \supset F(v_1))$, which is the sentence that Tarski used.]

The point, now, is that it is easy to get the Gödel number of $\exists v_1(v_1 = \overline{n} \wedge F(v_1))$ as an arithmetic function of the Gödel number of $F(v_1)$ and the number n, as we will soon see. We shall abbreviate $\exists v_1(v_1 = \overline{n} \wedge F(v_1))$ by $F[\overline{n}]$ (notice the square brackets!). To repeat an important point, the sentences $F[\overline{n}]$ and $F(\overline{n})$, though not the same, are equivalent (are both true or both false).

As a matter of fact, for any expression E, whether a formula or not, the expression $\exists v_1(v_1 = \overline{n} \wedge E)$ is a well-defined expression (though a meaningless one if E is not a formula), and we shall write $E[\overline{n}]$ as an abbreviation of (the possibly meaningless) expression $\exists v_1(v_1 = \overline{n} \wedge E)$. If E is a formula, so is $E[\overline{n}]$, and if E is a formula with v_1 as its only free variable, then $E[\overline{n}]$ is a sentence. But in all cases, $E[\overline{n}]$ is a well-defined expression.

We call a function $f(x, y)$ a $\sum_1$ function if the relation $f(x, y) = z$ is a $\sum_1$ relation. The following in the key lemma for Tarski's Theorem:

Lemma T_1. There is a $\sum_1$ function $r(x, y)$ such that for any expression E with Gödel number e, and any number n, $r(e, n)$ is the Gödel number of $E[\overline{n}]$.

Problem 11. Prove the above lemma. [Hint: The expression $\exists v_1(v_1 = \overline{n} \wedge E)$ consists of the expression $\exists v_1(v_1 =$ followed by $\overline{n}$, followed by the conjunction symbol $\wedge$, followed by E, followed by the right parenthesis. Consider the Gödel numbers of each of these parts.]

Now that we know that the relation $r(x, y) = z$ is arithmetic, Tarski's Theorem can be proved by essentially the same argument as that of Chapter 10. As in that chapter, call a sentence S_b with Gödel number b a *Gödel sentence* for a number set A if S_b is true iff $b \in A$. We must show that every arithmetic set A has a Gödel sentence.

Since the relation $r(x, y) = z$ is arithmetic, so is the relation $r(x, x) = y$ (as a relation between x and y). Let us define x^* to be $r(x, x)$. Thus the relation $x^* = y$ is arithmetic. A formula $F(v_1)$ with Gödel number b will be written $F_b(v_1)$. We note that b^* (which is $r(b, b)$) is the Gödel number of the sentence $F_b[b]$ (which can aptly be called the *diagonalization* of the formula $F_b(v_1)$). For any number set A, define $A^\#$ as the set of all numbers n such that $n^* \in A$. To show that every arithmetic set has a Gödel sentence, suppose A is arithmetic and let $F(x)$ be the formula that expresses A. Then $A^\#$ is also arithmetic, since $x \in A^\#$ iff $\exists y(x^* = y \wedge F(y))$. Let $F_b(v_1)$ be a formula (whose Gödel number is b) that expresses the set $A^\#$. Thus, for any number n, the sentence $F_b[\overline{n}]$ is true iff $n \in A^\#$, iff $n^* \in A$. Thus $F_b[\overline{n}]$ is true iff $n^* \in A$. In particular, taking b for n, $F_b[\overline{b}]$ is true iff $b^* \in A$, but b^* is the Gödel number of $F_b[\overline{b}]$, and so $F_b[\overline{b}]$ is a Gödel sentence for A. This proves that every arithmetic set A has a Gödel sentence.

Now, let T be the set of true sentences of arithmetic and let T_0 be the set of its Gödel numbers. The complement $\widetilde{T_0}$ of T_0 (which we recall is the set of all numbers not in T_0) cannot have a Gödel sentence, for such a sentence would be true iff its Gödel number is in $\widetilde{T_0}$, which would mean that the sentence is true iff its Gödel number was *not* the Gödel number of a true sentence, which is absurd. Thus $\widetilde{T_0}$ has no Gödel sentence. Since every arithmetic set *does* have a Gödel sentence, then $\widetilde{T_0}$ cannot be arithmetic. Hence T_0 cannot be arithmetic (for if it were expressed by some formula $F(x)$, the set $\widetilde{T_0}$ would be expressed by the formula $\sim F(x)$).

We have thus proved:

Theorem T_1 [Tarki's Theorem for Arithmetic]. The set of Gödel numbers of the true sentences is not arithmetic (not expressible by any formula).

Remark. We have proved Tarski's theorem for one particular Gödel numbering, the dyadic Gödel numbering, which is all we need for the semantic proof of Gödel's Theorem. Actually, Tarski's Theorem hold for any Gödel numbering having the property that there is an arithmetic relation $C(x, y, z)$ which reflects the concatenation relation of expressions, i.e. which is such that for any expressions X and Y, with respective Gödel numbers x and y, the relation $C(x, y, z)$ holds if and only if z is the Gödel number of the expression XY (X followed by Y). I give a proof of this in my book *The Gödelian Puzzle Book* [2013].

We know that if A is arithmetic, so is $A^\#$. We will later need to know that if A is $\sum_1$, so is $A^\#$.

Problem 12. Prove that if A is $\sum_1$, so is $A^\#$. [Note that if $R(x, y, z_1, \ldots, z_n)$ is $\sum_0$, then the relation $\exists x \exists y R(x, y, z_1, \ldots, z_n)$ is $\sum_1$.]

Problem 13. Prove that if A and B are $\sum_1$ sets, so are the union $A \cup B$ and the intersection $A \cap B$.

For any set W of sentences of first-order arithmetic, we take W_0 to be the set of dyadic Gödel numbers of the elements of W. We let T be the set of true sentences of first-order arithmetic. Tarski's Theorem, that T_0 is not arithmetic, implies that for any set W of true sentences (any subset of T), if W_0 is arithmetic, then W cannot be the whole of T (since W_0 cannot be the whole of T_0), and thus there must be a true sentence that is not in W. This is particularly significant when W is the set of all provable sentences of some axiom system for arithmetic that is correct (in the sense that all the provable sentences of the system are true), for then it follows that there is a true sentence that is not provable in the system, thus showing Gödel's result that the system is not complete.

Moreover, given a set W of true sentences such that W_0 is arithmetic, and given a formula that expresses the set W_0, we can effectively *find* a true sentence that is not in W!

Problem 14. Explain how.

In Chapter 13, we will consider the famous axiom system for Elementary (First-Order) Arithmetic known as Peano Arithmetic, The axioms introduced there are equivalent to those of several axiom systems that were developed for and focused on in the analysis of First-Order Arithmetic during the twentieth century. In that chapter, from the formula we get to express the set of Gödel numbers of the provable sentences of the system and as a consequence of the results of our hard work in this chapter, we will be able to explicitly exhibit a true sentence that is not provable in Peano Arithmetic.

* * *

We close the chapter with an important question: Is there a purely mechanical procedure by which we can determine which sentences of elementary arithmetic are true and which ones are false? To answer such questions, one must give a precise definition of a mechanical procedure. This bring us to the subject of *decision theory*, or *computability theory*, also called *recursion theory*, to which we turn in the next chapter.

Solutions to the Problems of Chapter 11

1. Proof that the following are arithmetic:
   (a) $x \, div \, y$ (x divides y) iff $\exists z(x \times z = y)$.
   (b) $x \, pdiv \, y$ (x divides y, and $x \neq 1$ and $x \neq y$) iff $x \, div \, y \wedge \sim(x = 1) \wedge \sim(x = y)$.

(c) $x < y$ (x is less than y) iff $x \leq y \wedge \sim(x = y)$.
(d) $prm\, x$ (x is a prime number) iff $\sim\exists y(y\, pdiv\, x)$.

2. Of course it is, since the relation $x \leq y$ is itself expressible in terms of plus and times: $x \leq y$ iff $\exists z(x + z = y)$.
3. We are to show that $x = 2^{L(y)}$, where $y \neq 0$ iff conditions C_1, C_2 and C_3 all hold.

   (a) Suppose that $x = 2^{L(y)}$, where $y \neq 0$. C_3 and C_2 clearly hold. Now let $r = L(y)$. Thus $x = 2^r$. Since r is the length of y, then by Condition (*), we have that $2^r - 1 \leq y \leq 2 \times (2^r - 1)$. Since $x = 2^r$, we have $x - 1 \leq y \leq 2(x - 1)$, which is C_1.

   (b) Conversely, suppose C_1, C_2 and C_3 all hold. By C_3, y is a natural number with value at least one, so has a dyadic numeral correspondent, making the length of y as a dyadic numeral, $L(y)$, a meaningful number, indeed a number with value at least one (which in turn, means the smallest possible value of x is 2). By C_2, $x = 2^r$ for some r. By C_1, $x - 1 \leq y \leq 2(x - 1)$, and so $2^r - 1 \leq y \leq 2 \times (2^r - 1)$. Then by (*), r is the length of y, i.e. $r = L(y)$. Since $x = 2^r$ and $r = L(y)$, then $x = 2^{L(y)}$.
4. To show that C_1 is $\sum_0$, we need to rewrite the expression we currently have for it, namely: $x - 1 \leq y \leq 2 \times (x - 1)$. This is because as it is it doesn't conform to the conditions on $\sum_0$ formulas. For instance we have to write the two inequalities separately, since compound inequalities are not allowed in $\sum_0$ formulas. Moreover, we have to get rid of the mathematical operation of subtraction, for which there is no symbol in First-Order Arithmetic. We can remedy those two problems by rewriting our original expression as: $(\exists z \leq x)(x = z + 1 \wedge z \leq y \wedge y \leq 2 \times z)$.
5. Solutions to the two parts:
   (a) $Pow_2(x)$ iff $(\forall y \leq x)[(y\; div\; x \wedge y \neq 1) \supset 2\; div\; y]$.
   (b) By (a), Condition C_2 is $\sum_0$ (i.e. constructive arithmetic) and we have proved that C_1 is $\sum_0$, and condition C_3 is obviously $\sum_0$; hence their conjunction is $\sum_0$, and this conjunction is equivalent to $x = 2^{L(y)}$. Thus the relation $x = 2^{L(y)}$ is $\sum_0$. Now, $x{*}y = x \times 2^{L(y)} + y$, and so $x{*}y = z$ is both meaningful and true iff the following formula is true:

   $$x \neq 0 \wedge y \neq 0 \wedge (\exists w \leq z)(w = 2^{L(y)} \wedge (x \times w) + y = z).$$

6. Solutions to the two parts:
   (a) $0 \neq x_1 \wedge 0 \neq x_2 \wedge 0 \neq x_3 \wedge x_1{*}x_2{*}x_3 = y$ is $\sum_0$, because it is equivalent to $0 \neq x_1 \wedge 0 \neq x_2 \wedge 0 \neq x_3 \wedge (\exists z \leq y)(x_1{*}x_2 = z \wedge z{*}x_3 = y)$.
   (b) We start the induction with $n = 2$. We already know that the relation $x_1{*}x_2 = y$ is $\sum_0$. Now suppose n is a number greater than or equal to 2 such that the relation $0 \neq x_1 \wedge \cdots \wedge 0 \neq x_n \wedge x_1 x_2{*} \cdots {*}x_n = y$ is $\sum_0$. Then $n + 1$ also has that property, since $x_1{*} \cdots {*}x_n{*}x_{n+1} = y$ iff $0 \neq x_1 \wedge \cdots \wedge 0 \neq x_n \wedge (\exists z \leq y)(x_1{*} \cdots {*}x_n = z \wedge z{*}x_{n+1} = y)$.

7. Solutions to the three parts:
   (a) xBy (x begins y) iff $x = y \vee (\exists z \leq y)(x{*}z = y)$.
   (b) xEy (x ends y) iff $x = y \vee (\exists z \leq y)(z{*}x = y)$.
   (c) xPy (x is a part of y) iff $xBy \vee xEy \vee (\exists z_1 \leq y)(\exists z_2 \leq y)$
   $(z_1{*}x{*}z_2 = y)$.
8. Take $K_0(x, y, z)$ to be the relation $K_1(x+1, y+1, z)$. This relation $K_0(x, y, z)$ is $\sum_0$, because it is equivalent to the $\sum_0$ relation

$$(\exists x_1 \leq z)(\exists x_2 \leq z)(x_1 = x+1 \wedge x_2 = y+1 \wedge K_1(x_1, x_2, z)).$$

   We must show that the relation $K_0(x, y, z)$ works.

   Let S be a finite sequence of ordered pairs of *natural* numbers. Let S' be the sequence of all ordered pairs $(x+1, y+1)$ such that $(x, y) \in S$. Applying Lemma K_1 to the sequence S', there is a number z such that for any natural numbers x and y, the pair $(x+1, y+1)$ of *positive* integers is in S' iff $K_1(x+1, y+1, z)$. Thus for any natural numbers x and y, the pair (x, y) is in S iff $(x+1, y+1)$ is in S', which is true iff $K_1(x+1, y+1, z)$, which is true iff $K_0(x, y, z)$. Thus $(x, y) \in S$ iff $K_0(x, y, z)$.
9. We first note that for any pair of natural numbers n and m for which $g(\overline{n}) = m$, the following is a sequence S containing $n+1$ ordered pairs of natural numbers such that if $(a, b) \in S$, then $g(\overline{a}) = b$, and if $(c+1, d) \in S$, then $(c, e) \in S$ for some natural number e:

$$S\text{: } (0, 12), (1, 12122), (2, 12122122), (3, 12122122122), \ldots, (n, m).$$

   Before continuing with the solution, let's pause and consider what the sequence just defined really consists in. The only binary relations we can express in Elementary Arithmetic are ones between pairs of natural numbers (i.e. relations $R(x, y)$, where, if a formula $F(x, y)$ of Elementary Arithmetic expresses the relation $R(x, y)$, it has to be the case that $R(m, n)$ is true for two natural numbers m and n iff $F(\overline{m}, \overline{n})$ is true by the definition of truth of formula given early in this chapter).

   But the Gödel numbering function g that we have defined is a function on *expressions* of Elementary Arithmetic, not natural numbers. By the definition of arithmeticity, we can't show that the function $g(X) = m$ is arithmetic, for X an arbitrary expression, and m its Gödel number, because all relations in First-Order Arithmetic have as their domains the natural numbers. But in the relation (expressed in ordinary mathematical English as) $g(\overline{n}) = m$, we have in $g(\overline{n})$ what's called the *composition* of two functions. The bar function, $\overline{n}$, is a function on the natural numbers that yields the *expression* of Elementary Arithmetic that is a zero followed by n accents. $g(\overline{n})$ is thus a functional expression that says to first do what the bar function says to n and then apply the Gödel numbering function g to the resulting expression (to yield a natural number).

So the $g(\overline{n}) = m$ relation we need to express is indeed a relation between two natural numbers n and m.

And to express that relation, we're going to make use of a sequence of ordered pairs of natural numbers. Here is the sequence again:

$$(0, 12), (1, 12122), (2, 12122122), (3, 12122122122), \ldots, (n, m).$$

But first let's be very clear about this sequence. It is meant to contain the first n ordered pairs (n, m) satisfying the relation between natural numbers $g(\overline{n}) = m$ that we're interested in. It's giving the ordered pairs reflected in the following table (where I indicate the notation used for the natural numbers involved in the various columns by $Dec(n)$ stands for "the decimal notation for n", $P(n)$ stands for "the Peano numeral designating n," and $Dy(g(P(n)))$ stands for "the dyadic numeral corresponding to the Gödel number of the Peano numeral designating n"):

n		m, or $g(\overline{n})$
$Dec(n)$	$P(n)$	$Dy(g(P(n)))$
0	0	12
1	$0'$	12122
2	$0''$	12122122
3	$0'''$	12122122122
4	$0''''$	12122122122122

It may seem strange that I've written the first of the two numbers in each ordered pair in the sequence in decimal notation, and the second number in Gödel notation, but realize that both n and m are representations of the same number. In this case, we are *using* the difference in the representations. How are we doing this, you may ask. To help see this, let's expand the above table:

	n			m, or $g(\overline{n})$
$Dy(n)$	$Dec(n)$	$P(n)$	$Dy(g(P(n)))$	$Dec(g(P(n)))$
none	0	0	12	4
1	1	$0'$	12122	42
2	2	$0''$	12122122	346
11	3	$0'''$	12122122122	3802
12	4	$0''''$	12122122122122	22234

From this table, we see that if we wanted to keep the same notation in the way we specify our ordered pairs in the sequence defined above, we couldn't quite do it with dyadic notation, because it doesn't have a numeral designating zero.

But the table also shows us that if we wanted to use decimal notation for both numbers in the ordered pairs showing us the value of our function for our first five integers, and we abbreviate the composite function $g(\overline{n})$ by $g'(n)$, expressing both the argument and value of g' in decimal notation, we have, for the first 5 natural numbers, $g'(0) = 4$, $g'(1) = 42$, $g'(2) = 346$, $g'(3) = 3802$, $g'(4) = 22234$. And we could write the sequence we started with for $n = 4$ with all the numbers in decimal notation as follows:

$$(0, 4), (1, 42), (2, 346), (3, 3802), (4, 22234).$$

In any case, in terms of a relation between natural number *expressed in decimal notation*, those *are* the ordered pairs corresponding to our function $g(\overline{n}) = m$ for the first 5 natural numbers. But I hope you can see why, in this particular case, it is convenient to write the numbers that come first in the ordered pairs of the sequence in decimal notation, and to write the numbers that come second in the ordered pair in Gödel notation.

Now we need to see why understanding the existence of such a sequence of ordered pairs help us prove that $g(\overline{n}) = m$ is not only arithmetic but $\sum_1$.

First let's note that by the construction of the sequence we described, for any numbers n and m, $g(\overline{n}) = m$ iff there is a finite sequence S of ordered pairs such that:

(1) (n, m) is an ordered pair in the sequence S.
(2) For every pair (a, b) in S, either $(a, b) = (0, 12)$, or there is some pair (c, d) in S such that $(a, b) = (c + 1, d{*}122)$.

Our construction has shown that *if* $g(\overline{n}) = m$, there does exist a finite sequence S such that (1) and (2) hold.

Conversely, suppose S is a finite sequence such that (1) and (2) hold. From (2), it follows by mathematical induction on a that for every pair (a, b) in S, $g(\overline{a}) = b$. In particular, by (1), we see that $g(\overline{n}) = m$.

It should be clear now that $g(\overline{u}) = v$ iff (u, v) is any member of any sequence satisfying (1) and (2) for some pair of natural numbers n, m.

It follows from this and Lemma K_0 that the following $\sum_1$ condition expresses $g(\overline{x}) = y$:

$$\begin{aligned}&\exists z[K_0(x, y, z) \wedge (\forall v \leq z)(\forall w \leq z)(K_0(v, w, z)\\&\quad \supset [(v = 0 \wedge w = \overline{4})\\&\qquad \vee (\exists v_1 \leq z)(\exists w_1 \leq z)(K_0(v_1, w_1, z) \wedge v = v_1 + 1 \wedge w = w_1{*}122)])]\end{aligned}$$

Or, equivalently:

$$\begin{aligned}&\exists z[K_0(x, y, z) \wedge (\forall v \leq z)(\forall w \leq z)(K_0(v, w, z)\\&\quad \supset [\sim(v = 0 \wedge w = \overline{4})\\&\qquad \supset (\exists v_1 \leq z)(\exists w_1 \leq z)(K_0(v_1, w_1, z) \wedge v = v_1 + 1 \wedge w = w_1{*}122)])]\end{aligned}$$

Note 1: The formula that defines $K_0(x, y, z)$ that we have to substitute in here twice contains all the conditions that guarantee that z is a code number for a finite sequence of ordered pairs of natural numbers. All we had to add here were conditions that guaranteed it is the kind of sequence we are using for the solution to the problem. Remember of course that $\overline{4}$ is really an abbreviation for the symbol 0 followed by four accents.

Note 2: Perhaps some readers remember that $K_0(x, y, z)$ is an abbreviation for $K_1(x+1, y+1, z)$ and worry whether the subformula $(v = 0 \wedge w = \overline{4})$ shouldn't really be $(v = 0' \wedge w = \overline{42})$. Let me show you why that isn't the case. If we substitute K_1 (with the appropriate arguments) for the K_0's that occur in the preceding formula, we get:

$$\exists z[K_1(x+1, y+1, z) \wedge (\forall v \leq z)(\forall w \leq z)(K_1(v+1, w+1, z)$$
$$\supset [(v = 0 \wedge w = \overline{4}) \vee (\exists v_1 \leq z)(\exists w_1 \leq z)(K_1(v_1+1, w_1+1, z)$$
$$\wedge\, v = v_1 + 1 \wedge w = w_1{*}122)])]$$

You should have understood everything inside the quantification on z and after the $K_0(x, y, z)$ *in the original formula* as saying, "for every (v, w) also in the sequence encoded in z, if $(v, w) \neq (0, \overline{4})$ [i.e. if, as we stated it in our metalanguage, using dyadic notation for the second number in the ordered pair, $(v, w) \neq (0, 12)$], then, the next lower ordered pair (in the order we put on our ordered pairs!) is also in the sequence. For example, if (2, 12122122) is in the sequence, so is (1, 12122). The way to see we did the right thing by using $(v = 0 \wedge w = \overline{4})$ is to look carefully at our formula written with K_1 instead of K. Let's just look at this part:

$$(\forall v \leq z)(\forall w \leq z)(K_1(v+1, w+1, z) \supset \ldots$$
$$(\exists v_1 \leq z)(\exists w_1 \leq z)(K_1(v_1+1, w_1+1, z) \wedge v = v_1 + 1 \wedge w = w_1{*}122))$$

Letting S be the sequence the full statement claims to exist [which we want to have the form (0, 12), (1, 12122), (2, 12122122), ...] we know that $K_1(x+1, y+1, z)$ says that $(x, y) \in S$. Thus, when we notice that $v = v_1 + 1$ implies $v_1 = v - 1$ and that $w = w_1{*}122$ implies that w_1 is the Gödel number of the positive integer one smaller than the positive integer for which w is the Gödel number, then we see that

(1) $K_1(v+1, w+1, z)$ says that $(v, w) \in S$;
(2) $K_1(v_1+1, w_1+1, z)$ says that $(v-1, w_1) \in S$.

And the only (v, w) in our sequence S for which the second claim isn't true is when $(v, w) = (0, \overline{4})$, because zero is the only natural number for which subtracting 1 is meaningless (in the domain of natural numbers).

There is one thing that those of you who saw this from the beginning may have learned from this exercise, though, namely that if we hadn't gone to K_0

from K_1, we could have eliminated one variable in our full formula expressing $g(\overline{x}) = y$:

$$\exists z[K_1(x+1, y+1, z) \wedge (\forall v \leq z)(\forall w \leq z)(K_1(v+1, w+1, z) \supset [(v = 0 \wedge w = \overline{4}) \vee (\exists w_1 \leq z)(K_1(v, w_1+1, z) \wedge w = w_1{*}122)])].$$

10. Solution to the two parts:
    (a) First let us note that for any relation $R(x, y, z_1, \ldots, z_n)$, the following two conditions (as relations between $z_1, \ldots, z_n$) are equivalent:

    (1) $\exists x \exists y R(x, y, z_1, \ldots, z_n)$.
    (2) $\exists w(\exists x \leq w)(\exists y \leq w)\ R(x, y, z_1, \ldots, z_n)$.

    It is obvious that (2) implies (1). Now suppose that $z_1, \ldots, z_n$ are numbers such that (1) holds. Then there are numbers x and y such that $R(x, y, z_1, \ldots, z_n)$ holds. Let w be the maximum of x and y. Then $(\exists x \leq w)(\exists y \leq w)\ R(x, y, z_1, \ldots, z_n)$ holds for such a number w, and so (2) holds.

    Suppose now that $R(x, y, z_1, \ldots, z_n)$ is $\sum_0$. Let w be a variable that does not occur free in $R(x, y, z_1, \ldots, z_n)$. Then the relation $(\exists x \leq w)(\exists y \leq w)\ R(x, y, z_1, \ldots, z_n)$ (as a relation between w, $z_1, \ldots, z_n$) is also $\sum_0$; hence the relation (2) is $\sum_1$; hence (1), being equivalent to (2), is also $\sum_1$.

    This proves that if $R(x, y, z_1, \ldots, z_n)$ is $\sum_0$, then the relation $\exists x \exists y R(x, y, z_1, \ldots, z_n)$ (as a relation between $z_1, \ldots, z_n$) is $\sum_1$.

    (b) If follows from (a) that if a relation $R(x, z_1, \ldots, z_n)$ is $\sum_1$, so is the relation $\exists x R(x, z_1, \ldots, z_n)$ (as a relation between $z_1, \ldots, z_n$). (Reason: Suppose $R(x, z_1, \ldots, z_n)$ is $\sum_1$. Then there is a $\sum_0$ relation $S(x, x_1, \ldots, x_n, y)$ such that $R(x, z_1, \ldots, z_n)$ iff $\exists y S(x, x_1, \ldots, x_n, y)$, hence $\exists x R(x, z_1, \ldots, z_n)$ iff $\exists x \exists y S(x, x_1, \ldots, x_n, y)$, which is $\sum_1$ by (a).

11. Let a be the Gödel number of the expression "$\exists v_1(v_1 =$" (which you can write down explicitly, if desired). The Gödel number of $\overline{n}$ is $g(\overline{n})$. The Gödel number of "$\wedge$" is g_8. Let e be the Gödel number of E. The Gödel number of the right parenthesis is g_4. Hence the Gödel number of the whole expression $\exists v_1(v_1 = \overline{n} \wedge \mathrm{E})$ is $a{*}g(\overline{n}){*}g_8{*}e{*}g_4$. We thus take $r(x, y)$ to be $a{*}g(\overline{y}){*}g_8{*}x{*}g_4$.

    Now we must verify that the relation $r(x, y) = z$ is $\sum_1$. Well $r(x, y) = z$ is the relation

    (1) $a{*}g(\overline{y}){*}g_8{*}x{*}g_4 = z$, which is equivalent to
    (2) $\exists w(w = g(\overline{y}) \wedge a{*}w{*}g_8{*}x{*}g_4 = z)$.

    Now, the relation $w = g(\overline{y})$ has been shown to be $\sum_1$, and so there is $\sum_0$ relation $S(w, y, v)$ such that $w = g(\overline{y})$ iff $\exists v S(w, y, v)$. Hence (2) is equivalent to:

(3) $\exists w \exists v(S(w, y, v) \wedge a{*}w{*}g_8{*}x{*}g_4 = z)$.

Since the relation $S(w, y, v) \wedge a{*}w{*}g_8{*}x{*}g_4 = z$ is $\sum_0$, then (3) is $\sum_1$ by Problem 10.

12. By definition of $A^{\#}$, $x \in A^{\#}$ iff $r(x, x) \in A$. Also, $r(x, x) \in A$ iff

(1) $\exists y(r(x, x) = y \wedge y \in A)$.

(Here $y \in A$ is not, strictly speaking, a formula of Elementary Arithmetic. But see below how it can be taken to be an abbreviation for an arithmetic formula, indeed of a $\sum_1$ formula, since A is assumed to be $\sum_1$ in the statement of this problem.)

Now suppose that A is $\sum_1$. We will first show that $r(x, x) = y \wedge y \in A$ is $\sum_1$.

Since A is $\sum_1$, there is a $\sum_0$ relation $R(y, z)$ such that $y \in A$ iff $\exists z R(y, z)$.

Since the relation $r(x, y) = z$ has been shown to be $\sum_1$, then the relation $r(x, x) = y$ is $\sum_1$, and so there is a $\sum_0$ relation $S(x, y, w)$ such that $r(x, x) = y$ iff $\exists w S(x, y, w)$. Therefore $r(x, x) = y \wedge y \in W$ iff $\exists w \exists z(S(x, y, w) \wedge R(y, z))$. This relation is $\sum_1$ by Problem 10; since the relation $S(x, y, w) \wedge R(y, z)$ is $\sum_0$. Thus the relation $r(x, x) = y \wedge y \in A$ is $\sum_1$, hence so is the condition $\exists y(r(x, x) = y \wedge y \in A)$ (again by Problem 10). Thus $A^{\#}$ is $\sum_1$.

13. Suppose A and B are both $\sum_1$. Then there are $\sum_0$ relations $R_1(x, y)$, $R_2(x, y)$ such that $x \in A$ iff $\exists y R_1(x, y)$, and $x \in B$ iff $\exists y R_2(x, y)$. Then $x \in A \cup B$ iff $\exists w(\exists y \leq w)(\exists z \leq w)(R_1(x, y) \vee R_2(x, z))$ is true. Also, $x \in A \cap B$ iff $\exists w(\exists y \leq w)(\exists z \leq w)(R_1(x, y) \wedge R_2(x, z))$.

14. To begin with, our proof that for any arithmetic set A there is a Gödel sentence for A was wholly constructive, in the sense that given an arithmetic formula $F(y)$ that expresses the set A, we can effectively exhibit a Gödel sentence for A. We do this as follows:

We must first exhibit a formula $H(v_1)$ that expresses the set $A^{\#}$. Well, $x \in A^{\#}$ iff $\exists y(r(x, x) = y \wedge y \in A)$. We thus take a formula $\varphi(x, y)$ that expressed the relation $r(x, x) = y$. We then take $H(v_1)$ to be the formula $\exists y(\varphi(v_1, y) \wedge F(y))$. Then $H(v_1)$ expresses the set $A^{\#}$. We let h be the Gödel number of $H(v_1)$ and $H[\overline{h}]$ is then a Gödel sentence for A, since $H[\overline{h}]$ is true iff $h \in A^{\#}$, which is true iff $r(h, h) \in A$, and $r(h, h)$ is the Gödel number of $H[\overline{h}]$.

Now suppose that W is a set of true sentences and that W_0 is arithmetic, and that $F(y)$ is an arithmetic formula that expresses W_0. Its negation $\sim F(y)$ then expresses the complement $\widetilde{W_0}$ of W_0. From the formula $\sim F(x)$ we can then find a Gödel sentence X for $\widetilde{W_0}$, as explained above. Thus X is true iff its Gödel number $X_0 \in \widetilde{W_0}$, which is true iff $X_0 \notin W_0$, which is true iff $X \notin W$. Thus X is true iff X is not in W; and since W is a subset of the set of true sentences, X must be true but not in W.

12
Formal Systems

Just what is meant by a purely *mechanical* procedure? Informally, a mechanical method is one that can be carried out without any creative ingenuity – one that can be carried out by a computer. But this informal characterization is not good enough for mathematical purposes, so how can we give a precise definition of what it means for a procedure to be "mechanical"? In the twentieth century about a dozen or so workers in mathematical logic and computer science independently gave their definitions (for example, the Recursive Functions of Kurt Gödel and Jacques Herbrand, the Machines of Alan Turing and the related Register Machines, the Lambda Definability of Alonzo Church, the Combinatory Logic of Moses Schönfinkel, the Canonical Systems of Email Post, the Elementary Formal Systems of Raymond Smullyan). The interesting thing is that all these definitions, which appeared to be quite different on the surface, turned out to be equivalent! Any process that is "mechanical" according to any one of the definitions is also mechanical according to any other of the definitions. This constitutes strong heuristic evidence that these definitions have each correctly captured what is meant by a "mechanical" procedure.

The concept of a mechanical procedure is closely related to the notion of a *formal* mathematical system. Speaking informally for the moment, a *formal proof system for mathematical logic* – in the future here, just a *formal system* – is one whose provable sentences can be generated by a purely mechanical procedure. Going in the other direction, once we have defined what it means for a system to be *formal*, we can define a set to be generated by a mechanical procedure if the set is definable in some formal system. This is the approach we will take. We first define a very simple type of formal system that I call an *elementary formal system*, and then define formal systems and mechanical operations in terms of these elementary formal systems. The whole subject of recursion theory (also called computability theory) can be simply and elegantly developed in the context of elementary formal systems [Smullyan, 1961].

Elementary Formal Systems

Before giving a precise definition of an elementary formal system, an informal discussion should be helpful.

Elementary formal systems provide a means of converting *implicit* definitions into *explicit* ones. Let me explain what I mean by that: By an *implicit* or "recursive" definition I mean one of the following type, which is frequent in the mathematical literature: Instead of defining as set or relation W outright, W is *implicitly* defined by giving a set of rules for membership in W – rules of the form "such and such is in W", or if such and such is in W, so is so-an-do." [For example, the definition of a *formula* of Propositional Logic: Propositions are formulas; if X and Y are formulas, so are $\sim X$, $X \supset Y$, $X \wedge Y$, $X \vee Y$.] Then comes the final clause (the so-called *recursion* clause): "Nothing is in W unless its being so is a consequence of the above rules." Consequence in what logic? Elementary formal systems, soon to be defined, provide one such type of logic.

For the moment, it should be helpful to think of elementary formal systems as *programs* for generating sets and relations. Let us consider some examples.

Let us consider just two symbols a and b and the set of all strings (expressions) in just the symbols a and b. For any two such strings x and y, by xy will be meant x followed by y, i.e. the concatenation of x and y.

Now, suppose we wish to generate the set A of all alternating strings, i.e. strings which contain no consecutive occurrences of a or of b. The following facts clearly hold, and are being used here to give an implicit definition of what it means for a string to be in the set of alternating strings:

(1) a is alternating.
(2) b is alternating.
(3) ab is alternating.
(4) ba is alternating.
(5) If xa is alternating, then so is xab.
(6) If xb is alternating, then so is xba.

Also, no string is alternating unless its being so is a consequence of the above six conditions.

We can now give the following instructions for generating the set A:

(1) Put a in A.
(2) Put b in A.
(3) Put ab in A.
(4) Put ba in A.
(5) For any x, if xa is in A, put xab in A.
(6) For any x, if xb is in A, put xba in A.

Now, computer programs are written in *symbolic* languages, i.e. with just one interpretation possible for each string of symbols representing a command occurring in them (true even when the programs are written in something very close to English, which occurs with some programming languages these days). Thus, to emphasize that we are aiming for elementary formal systems that can be seen as *determinate programs* for a computer (i.e. ones that will work the same way every time they receive the same input), we will abbreviate the command "put x in A" with the string of symbols "Ax". Thus 1, 2, 3, 4 are now written

(1) Aa.
(2) Ab.
(3) Aab.
(4) Aba.

These lines can be called "unconditional rules," for each just says to put a specific string of symbols into the set A. What remains from the previous way of specifying this example are the last two lines from the earlier version above. These are "conditional rules." For each of them says "if an element of a certain form is in A, put this other element (related to it) into A." The if-then relation will be symbolized here by "$\rightarrow$" (to avoid confusion with the implication sign "$\supset$" of propositional and First-Order Logic), and so (5) and (6) are written thus:

(5) $Axa \rightarrow Axab$.
(6) $Axb \rightarrow Axba$.

The letter "x" is used as a variable which stands for any string of a's and b's. By an *instance* of an expression involving the variable "x" is meant the result of substituting a string of a's and b's for all occurrences of x in an expression (the same string for all the occurrences of x).

Let us see how a computer could interpret these lines as specifying a mechanical procedure to generate the set A. By (1), the single letter a gets put in A. By (2), (3) and (4), the strings b, ab, and ba get put in A. Next, the computer takes the first element put into the set A, namely a, and looks at the conditional rules to see if either of them apply to it. Neither do, because a is a singleton and the x in each rule stands for a string in a's and b's of length at least one. And when the computer looks at b, the second element put into A, no rule applies to it either. So the computer then considers the third element put into A, namely ab. Rule (5) does not apply, but in (6), the computer can replace x by a, and it will have the instance $Aab \rightarrow Aaba$, which it interprets as, "If ab is in A, put aba in A." Since the computer knows ab is in A, it will put aba into A. Next it takes up the 4th element placed into A, namely ab, and this time Rule (5) applies but not Rule (6): in (5) the computer replaces x by ab, and has the instance $Aaba \rightarrow Aabab$, which it interprets as meaning, "If aba is in A, put $abab$ in A." Since it know aba is in A, it puts $abab$ into A. Continuing in this manner, i.e. taking into consideration the elements put into A in the order in which they were placed in it, the computer will generate every alternating string sooner or later.

So we now have written out 6 lines, each consisting of a string of symbols that we just saw could be interpreted as a line in a computer program. Very soon we will see that elementary formal systems can be much more complicated than this simple example, and to understand how they can be interpreted as a computer algorithm gets a little more complicated. For instance, it will be important to be precise about the order in which the computer should apply the rules of the elementary formal system in order to not get stuck repeating a single rule to the avoidance of other rules (say if $Axab \to Axabab$ happened to be an added rule here and the computer decided to use that rule repeatedly as often as it could be applied, the computer wouldn't generate all possible alternating strings).

Thus, the above is a simple example of an elementary formal system, more specifically an elementary formal system over the alphabet $\{a, b\}$. The symbol "A" is an example of what is called a *predicate*. Instead of looking at the above system as specifying a program for generating the set of alternating strings, we will now look at the system as a mathematical *axiom system*, in which the instances of (1), (2), (3), (4) are the axioms of the system, and (5) and (6) are inference rules. For instance (5), read as an inference rule, is the following, "From Axa, to infer $Axab$," and (6) would be "From Axb, to infer $Axba$." Then to say that "A" represents the set of all alternating strings is to say that for any string x in a and b, the sentence Ax is provable iff x is an alternating string.

Exercise. But how do we know that A really represents the set of alternating strings? We need to be able to see very clearly both that if x is alternating, then Ax really is provable, and also that if Ax is provable, then x really is alternating. See if you can prove these two things.

Elementary formal systems also provide means of representing *relations* between strings. For example, let K be the 3-symbol alphabet $\{a, b, c\}$. By the *reverse* of a string composed out of these three symbols is meant the string in which the symbols are written in reverse order. For example, the reverse of $cabbab$ is $babbac$. This relation of reversal is completely determined by the following conditions:

(1) The symbol a alone is the reverse of itself;
(2) The symbol b alone is the reverse of itself;
(3) The symbol c alone is the reverse of itself;
(4) If x is the reverse of y then ax is the reverse of ya;
(5) If x is the reverse of y then bx is the reverse of yb;
(6) If x is the reverse of y then cx is the reverse of yc.

We use the symbol "R" as a name of the reverse relation, and we wish to generate statements of the form Rx, y whenever x is the reverse of y, and never when x is not the reverse of y. The following instructions accomplish this:

(1) $Ra, a.$
(2) $Rb, b.$
(3) $Rc, c.$
(4) $Rx, y \to Rxa, ay.$

(5) $Rx, y \to Rxb, by$.
(6) $Rx, y \to Rxc, cy$.

This system can be shortened a bit by observing the following facts:

(1) a is a single symbol.
(2) b is a single symbol.
(3) c is a single symbol.
(4) If x is a single symbol, then x is its own reverse.
(5) If x is the reverse of y, and if z is a single symbol, then xz is the reverse of zy.

Thus, we abbreviate "x is a single symbol" by "Sx", and have the following system:

(1) Sa.
(2) Sb.
(3) Sc.
(4) $Sx \to Rx, x$.
(5) $Rx, y \to Sz \to Rxz, zy$.

We are here using the symbol "$\to$" as implication with *association to the right*, that is, for any statements X, Y and Z, the statement $X \to Y \to Z$ is read "if X is true, then Y implies Z" or as "if X is true, then if Y is true, so is Z," and is *not* read as "if X implies Y, then Z is true" (which would be association to the left). Similarly, for any four statement X, Y, Z and W, the statement $X \to Y \to Z \to W$ is read "if X is true, then if Y is true, then if Z is true, so is W", which means the same thing as "if X, Y and Z are all true, so is W", and could alternatively be symbolized: $(X \,\&\, Y \,\&\, Z) \to W$," where "&" is the symbol for "and." For technical reasons, it is best not to use the extra-logical connective "&," but only to use the logical connective "$\to$" for "implies."

Here is another system for the reverse relation, which is still shorter and also works.

(1) Ra, a.
(2) Rb, b.
(3) Rc, c.
(4) $Rx, y \to Rz, w \to Rxz, wy$.

Thus we have seen that there can be various elementary formal systems to generate the say set of relations, and some may generate more than one set and/or relation simultaneously.

Definition of an Elementary Formal System

Before preceding further, I should now give a precise definition of an *elementary formal system.*

By an *alphabet* K we shall mean a finite sequence of elements called the *symbols, signs*, or *letters* of K. Any finite sequence of symbols of K is called a *string*, or an *expression* or a *word* in K, of more briefly, a K-string. For any K-strings X and Y by XY is again meant the sequence X followed by the sequence Y. For example, if X is the string *am* and Y is the string *hjkd*, then XY is the string *amhjkd*. As usual, the string XY is called the *concatenation* of X and Y.

By an *elementary formal system* (E) over K we mean a collection of the following items:

(1) An alphabet K.
(2) Another alphabet of symbols called *variables*. We will usually use the letters x, y and z, with or without subscripts, as our variables.
(3) Still another alphabet of signs called *predicates*, each of which is assigned a positive integer called the *degree* of the predicate. We usually use capital letters for predicates.
(4) Two more symbols called the *punctuation sign* (usually a comma) and the *implication sign* (usually "$\rightarrow$").
(5) A finite sequence of strings which are formulas, according to the definition given below.

But first some preliminary definitions. By a *term* we mean any string composed of symbols of K and variables. For example, if a, b, c are symbols of K, and x, y are variables, then $aycxxbx$ is a term. A term without variables will be called a *constant term*. By an *atomic formula* we mean an expression Pt, where P is a predicate of degree 1 followed by a term t, or an expression Rt_1, t_2, where R is a predicate of degree 2 and t_1 and t_2 are terms, or more generally, for any positive integer n, a predicate of degree n followed by n terms separated by commas. If F and G are formulas, then $F \rightarrow G$ is a formula.

By a *sentence* we shall mean any formula without variables.

By an *instance* of a formula, we shall mean the result of substituting strings in K for occurrences of all the variables of the formula, with the understanding that if a variable has more than one occurrence in the formula, the same string in K must be substituted for each of the occurrences of the variable. For example, consider a formula $Paxbycx$, where a, b, c are symbols of K and x and y are variables. Suppose we substitute ab for x and ca for y. We then obtain the instance $Paabbcacab$. If a formula has no variables, i.e. if it is a sentence, its one and only instance is itself.

An elementary formal system (E) thus consists of a finite set of these formulas called the *initial formulas*, or *axioms schemes*, of the system. The set of all instances of all the axiom schemes are called the *axioms*, or *initial sentences*, of the system.

We now define a formula to be *provable* in the system if its being so is a consequence of the following two conditions:

(1) Every instance of an axiom scheme of the system is provable in the system.
(2) For every *atomic* sentence X and any sentence Y, if X is provable and $X \rightarrow Y$ is provable, then Y is provable.

More precisely, by a *proof* in the system (E) is meant a finite sequence of formulas of (E) (usually displayed vertically, rather than horizontally), called the *lines* of the proof, such that for each line Y of the proof, either Y is an instance of an axiom scheme of (E), or there is an *atomic sentence* X such that both X and $X \rightarrow Y$ are earlier lines of the proof (this rule of inference is called the *Rule of Detachment*). Note that no variables at all occur on any line of a proof, although conditional sentences (sentences containing $\rightarrow$) do occur.

A sentence X is then called *provable* in (E) if it is the last line of some proof, and such a proof is called a proof of X. Note that we are principally interested in the set of provable *atomic sentences* of the elementary formal system (E), for they are the ones that tell us the n-tuples of K-strings that are generated for each predicate of degree n that occurs in the rules of (E).

We said earlier that elementary formal systems are a way to explain the concept of a mechanical procedure. To understand this better, you might wish to see how to encode more complex mechanical procedures as elementary formal systems. It is beyond the scope of the book to give many realistic examples, but we will in fact give one very complex example soon, by encoding a logical system (Peano Arithmetic) in an elementary formal system. But in order to understand what it means that the specification of an arbitrary elementary formal system actually can be interpreted as a program to generate all the provable atomic sentences of the system, ideally the reader could try to think through in detail how the axiom schemes of an arbitrary elementary formal system could be interpreted as a computer program to successively generate all the provable atomic sentences (which are the provable formulas we are usually the most interested in). The understanding gained in Chapter 2 about how to enumerate (mechanically) all the elements of variously defined denumerable sets will help in this endeavor (for instance if an atomic formula that is an axiom scheme contains precisely n different variables, just to get all the instances of that single axiom scheme, we will have to enumerate all the n-tuples of K-strings, so as to use them to get all the provable atomic sentences that happen to be instances of that one axiom scheme).

Representability

For any elementary formal system (E) over an alphabet K, a predicate P of degree 1 is said to *represent* the set of all *constant terms* t such that Pt is provable in the system (E). A set S of constant terms of (E) is said to be *representable* in (E) if some predicate of (E) represents it.

A relation $R(x, y)$ will be said to be represented by a predicate P of degree 2 if for all constant terms t_1 and t_2 the relation $R(t_1, t_2)$ holds if and only if the sentence Pt_1, t_2 is provable in the system (E). More generally, a relation $R(x_1, \ldots, x_n)$ of n arguments will be said to be represented by a predicate P of degree n if for all constant terms $t_1, \ldots, t_n$, the relation $R(t_1, \ldots, t_n)$ holds if and only if the sentence $Pt_1, \ldots, t_n$ is provable in the system (E).

A set or relation is said to be *formally representable* over K, or K-representable, if it is represented in some elementary formal system over K. Finally, a set or relation is said to be *formally representable* if it is formally representable in some elementary formal system over some alphabet K.

Problem 1. Suppose W_1 is a set representable in an elementary formal system (E_1) over the alphabet K, and that W_2 is a set representable in an elementary formal system (E_2) over the same alphabet. Is there necessarily a single elementary formal system in which both W_1 and W_2 are representable?

For any set S of K-strings, S is called *solvable over* K if both S and its complement $\tilde{S}$ (the complement being with respect to the set of all K-strings) are both representable over K. A set will be called *solvable* if it is solvable over some alphabet K.

Discussion

The word *solvable* is well chosen. As mentioned earlier, given an elementary formal system (E), one can program a computer to generate the set of all sentences provable in (E). Now, suppose S is solvable over K. Then there are predicates P_1 and P_2 in an elementary formal system (E) such that for any string X of symbols of K, P_1X will be provable iff $X \in S$ and P_2X will be provable iff $X \notin S$. One then sets a computer going on the program that will generate all the members of the relations represented in (E), and, since the K-string X is either in S or it isn't, sooner or later either P_1X or P_2X will come out. If P_1X comes out, we will know that X is in S, whereas if P_2X comes out, we will know that X is not in S. Thus we have just seen that determining membership in S has been shown to be a mechanically solvable problem.

Now, suppose a set S is representable but not solvable, and assume S is represented in some elementary formal system by some predicate P. We would like to know of a given string X whether it is or is not a member of S. The best we can then do is to start a computer going to print out the provable sentences of the system. If X does belong to the set S, then sooner or later the computer will print out PX and we will know that X does belong to S. But if X doesn't belong to S, then the computer may run on forever, and at no stage can we know whether or not PX will be printed in the future. In short, if S is representable but not solvable, if X belongs to S, we will eventually know it, but if X doesn't belong to S, then we will never know it (unless by some creative ingenuity, we discover a way of finding out). Such a set S might aptly be called *semi-solvable.*

Do there exist sets which are semi-solvable, but not solvable? That is a fundamental question in recursion theory and will be answered later on.

Problem 2. Suppose W_1 and W_2 are sets that are both formally representable over K. Prove that their union $W_1 \cup W_2$ and their intersection $W_1 \cap W_2$ are both formally representable over K.

Problem 3. Suppose W_1 and W_2 are sets that are both *solvable* over K. Are $W_1 \cup W_2$ and $W_1 \cap W_2$ necessarily solvable over K?

Numerical Sets and Relations

I have found dyadic notation for the positive integers to be particularly useful for our present purposes. We let D be the two-sign alphabet $\{1, 2\}$, and an elementary formal system over D will be called an *elementary dyadic system.* Until further notice, we identify the positive integers with the dyadic numerals that denote them. A set or relation W will be called *dyadically enumerable* if it is representable over D. [It turns out that being dyadically enumerable is the same thing as being $\sum_1$, as defined in the last chapter! We will say more about this later.] We call A *dyadically solvable* if both A and $\widetilde{A}$ are dyadically enumerable.

Problem 4. Show that the following relations are dyadically enumerable.

(a) Sx, y (the successor of x is y)
(b) $x < y$
(c) $x = y$
(d) $x \leq y$
(e) $x \neq y$
(f) $x + y = z$
(g) $x \times y = z$
(h) $x^y = z$

A relation $R(x, y)$ is called *single-valued*, or *functional*, if for each x, there is one and only one y such that $R(x, y)$ holds. A relation $R(x, y, z)$ is called *single-valued* or *functional* if for every pair x and y, there is one and only one z such that $R(x, y, z)$ holds. [Similarly, a relation $R(x_1, \ldots, x_n, y)$ is called functional or single-valued if for every $x_1, \ldots, x_n$, there is one and only one y such that $R(x_1, \ldots, x_n, y)$ holds.]

Problem 5. Prove that if a relation $R(x, y)$ or $R(x, y, z)$ is single-valued, then if it is dyadically enumerable, it must be dyadically solvable.

Problem 6. Which of the relations of Problem 4 are dyadically solvable?

Arithmetization of Elementary Formal Systems

As in the last chapter, for any positive n, we let g_n be the number which in dyadic notation consists of 1 followed by n occurrences of 2 (e.g. $g_4 = 12222$). If K is any ordered alphabet $\langle a_1, a_2, \ldots, a_n\rangle$ of symbols, we assign dyadic Gödel numbers to all strings of symbols of K in the same way as we did in the last chapter for the ordered alphabet of elementary arithmetic; that is, for any string X of symbols of K, we take its dyadic Gödel number to be the results of replacing each occurrence of a_1 by g_1, a_2 by g_2, etc. (e.g. the Gödel number of $a_3a_1a_2$ is $g_3g_1g_2$, which is 122212122).

For any set W of strings in K, by W_0 we shall mean the set of (dyadic) Gödel numbers of the elements of W. The main purpose of this chapter is to prove that if W is formally representable over K, then the set W_0 is $\sum_1$. This has several important ramifications, as we will see.

Some preliminaries are in order. For any positive n, we let G_n be the set of all dyadic numerals compounded by concatenation from $g_1, \ldots, g_n$. [Thus for each $i \leq n$ $g_i \in G_n$, and for any X and Y in G_n, the numeral XY is in G_n.]

Problem 7. Prove that for any positive n, the set G_n is $\sum_0$.

Next, we need:

Substitution Lemma. Let L be an ordered alphabet $\langle k_1, \ldots, k_n, a_1, \ldots, a_m\rangle$, and let K be the ordered sub-alphabet $\langle k_1, \ldots, k_n\rangle$. For any string X compounded from symbols of L and variables $x_1, \ldots, x_t$ (which we assume are not symbols of L), let $I(X)$ be the set of all strings that result from substituting strings in K for all variables in X, and let $I_0(X)$ be the set of dyadic Gödel numbers of the strings in $I(X)$. Then the set $I_0(X)$ is $\sum_0$.

Problem 8. Prove the Substitution Lemma.

Now let us consider an elementary formal system (E) over an alphabet K. Let L be the set of symbols of K together with the predicates of (E) and the comma and the arrow (thus L is the alphabet out of which all sentences of (E) are built). We order L so that the symbols of K come at the beginning, and we let g be the dyadic Gödel numbering of all strings in L. We let Pr be the set of all provable sentences of (E).

We recall that by a *proof* in (E) is meant a finite sequence $X_1, \ldots, X_r$ of sentences of (E) such that for each $i \leq r$, the sentence X_i is either an instance of an axiom scheme of (E), or is derivable from two earlier members of the sequence by the rule of detachment (which we recall is that Y is directly derivable from X and $X \to Y$, providing that X is atomic).

We now need to assign to each finite sequence $X_1, \ldots, X_r$ of strings in L a number which we will call the *sequence number* of the sequence. We do this as follows:

Let $m = n+1$, where n is the number of symbols of L. As before, we let g_m be the dyadic numeral consisting of 1 followed by m occurrences of 2. For each term X_i of the sequence $X_1, \ldots, X_r$, we let a_i be the dyadic Gödel number of the term X_i. To the sequence $X_1, \ldots, X_r$ we assign the number $g_m a_1 g_m a_2 g_m \ldots g_m a_r g_m$.

We let $Seq(x)$ be the property that x is the sequence number of some sequence. We let $x \in y$ be the condition that y is a sequence number and x is the dyadic Gödel number of some term of the sequence. We let $pr(x, y, z)$ be the condition that z is the sequence number of some sequence Z, of which y is the Gödel number of some term Y and x is the Gödel number of some term X which occurs prior to the first (leftmost) occurrence of Y.

Problem 9. Prove that the conditions $Seq(x)$, $x \in y$ and $pr(x, y, z)$ are all $\sum_0$.

We now define $Der(x, y, z)$ to mean that x, y and z are respectively the Gödel numbers of expressions X, Y and Z of (E) such that Z is derivable from X and Y by the rule of detachment.

Problem 10. Prove that the relation $Der(x, y, z)$ is $\sum_0$.

Next, we define $Pf(x)$ to mean that x is the sequence number of a proof in (E), and we define $ypfx$ to mean that y is the sequence number of a proof in (E) of which x is the Gödel number of its last term.

Problem 11. Prove that the following conditions are all $\sum_1$:

(a) $Pf(x)$

(b) $ypfx$

(c) The set Pr of provable sentences of (E)

(d) For any representable set W, the set W_0 of dyadic Gödel numbers of the members of the set W is $\sum_1$.

We have now proved:

Theorem A. For any formally representable set W of expressions, the set W_0 of dyadic Gödel numbers of the members of W is $\sum_1$.

N.B. The proof of Theorem A that we have given is wholly constructive, in the sense that given an elementary formal system (E) in which W is represented, we can actually exhibit a $\sum_1$ formula that expresses the set W_0.

Of course, Theorem A holds not only for a set W, but also for any relation W, which the reader can easily verify.

Ramifications

As previously mentioned, the fact that for every formally representable set W, the set W_0 is $\sum_1$ has several important ramifications. For one thing, it, together with Tarski's Theorem yields:

Theorem T_1*. The set T of true sentences of first-order arithmetic is not formally representable.

Problem 12. Why?

This answers the question raised at the end of the last chapter, namely, is there a purely mechanical method of determining which sentences of first-order arithmetic are true and which ones are false? In terms of recursion theory, the question is whether or not the set T is solvable. Well, not only is it not solvable, but it is not even formally representable! Thus the set T cannot be generated by any purely mechanical device, let alone solved by one.

We will call an axiom system a *formal system* if the set of provable sentences of the system is formally representable.

By an axiom system for arithmetic, we shall mean an axiom system for first-order arithmetic, i.e. an axiom system whose formulas are those of first-order arithmetic. We are calling such a system *correct* if all the sentences provable in the system are true.

From Theorem T_1* immediately follows:

Theorem GT [After Gödel, Tarski]. Given any correct formal axiom system $\mathcal{S}$ for arithmetic, there is a true sentence not provable in $\mathcal{S}$.

N.B. For such a system $\mathcal{S}$, not only is it the case that there exists a true sentence not provable in $\mathcal{S}$, but given an elementary formal system (E) in which the set P of provable sentences is represented, we can actually exhibit a true sentence that is not provable in $\mathcal{S}$, for, as earlier remarked, from (E) we can exhibit a $\sum_1$ formula $F(x)$ that expresses the set P_0 of Gödel numbers of the provable sentences. The formula $F(x)$ is of course arithmetic. Hence by Problem 14 of the last chapter, taking F for W, we can find a true sentence not in P.

In what follows, $\mathcal{S}$ is a formal system of arithmetic, P is the set of provable sentences of $\mathcal{S}$, and P_0 is the set of dyadic Gödel numbers of the elements of P.

The proof of Theorem GT did not need the full force of the fact that P_0 is $\sum_1$ but required only the weaker fact that P_0 is arithmetic. However, for the Gödel proof based on omega consistency, and the Rosser proof, based on the weaker assumption of simple consistency, the fact that P_0 is not only arithmetic, but $\sum_1$ is needed.

We recall that a relation $R(x_1, \ldots, x_n)$ is said to be *definable* in a system $\mathcal{S}$ if there is a formula $F(x_1, \ldots, x_n)$, said to define R, such that for all numbers $a_1, \ldots, a_n$, the sentence $F(\overline{a_1}, \ldots, \overline{a_n})$ is provable if $R(a_1, \ldots, a_n)$ holds, and is refutable if $R(a_1, \ldots, a_n)$ doesn't hold. We say that $\mathcal{S}$ is $\sum_0$*-complete* if every $\sum_0$ relation is definable in $\mathcal{S}$. This is equivalent to the condition that every true $\sum_0$ sentence is provable in $\mathcal{S}$.

Theorem G [After Gödel]. For a formal and $\sum_0$-*complete* system $\mathcal{S}$, if the system is ω-consistent, then some sentence of $\mathcal{S}$ is undecidable.

Theorem R [After Rosser]. Suppose $\mathcal{S}$ is a formal $\sum_0$-*complete* system satisfying the additional condition that for any formula $F(x)$ and any number n,

L_1: If $F(\overline{0}), \ldots, F(\overline{n})$ are all provable, so is $\forall x(x \leq \overline{n} \supset F(x))$.
L_2: $\forall x(x \leq \overline{n} \vee \overline{n} \leq x)$ is provable.

Then, if the system is *simply consistent*, there is an undecidable sentence.

Problem 13. Prove Theorems G and R.

In the next chapter, we consider an actual axiom system for arithmetic, namely the famous system known as *Peano Arithmetic*, and show that it obeys the hypotheses of Theorems GT, G and R, thus completing the three famous incompleteness proofs for Peano Arithmetic.

Solutions to the Problems of Chapter 12

1. Let (E_1) be an elementary formal system over K in which W_1 is represented, and let (E_2) be an elementary formal system over K in which W_2 is represented. If there are any predicates common to (E_1) and (E_2), replace those predicates of (E_2) common to (E_1) by new symbols that are not in (E_1), and call the resulting system $(E_2)'$. Then take the axiom schemes of (E_1) together with those of $(E_2)'$ as axiom schemes of (E). Then W_1 and W_2 are both represented in (E).
2. Suppose W_1 and W_2 are both representable over K. By Problem 1, there is an elementary formal system (E) in which W_1 and W_2 are both represented, say by P_1 and P_2 respectively.
   (a) To represent $W_1 \cup W_2$, take a new predicate P and add the two axiom schemes:
   $P_1x \rightarrow Px$
   $P_2x \rightarrow Px$
   Then P represents $W_1 \cup W_2$.
   (b) To represent $W_1 \cap W_2$, again take a new predicate P and add the single axiom scheme:
   $P_1x \rightarrow P_2x \rightarrow Px$
   Then P represents $W_1 \cap W_2$.
3. Yes, they are: Suppose W_1 and W_2 are both *solvable* over K. Thus W_1, $\widetilde{W_1}$, W_2, $\widetilde{W_2}$, are each K-representable (formally representable over K). Since W_1 and W_2 are K-representable, then so are $W_1 \cup W_2$ and $W_1 \cap W_2$ (by Problem 2). We must show that their complements, $\widetilde{W_1 \cup W_2}$ and $\widetilde{W_1 \cap W_2}$ are K-representable. Well, since $\widetilde{W_1}$ and $\widetilde{W_2}$ are both K-representable, so is

$\widetilde{W_1} \cap \widetilde{W_2}$. But $\widetilde{W_1} \cap \widetilde{W_2} = \widetilde{W_1 \cup W_2}$. Thus $\widetilde{W_1 \cup W_2}$ is K-representable. The case of $\widetilde{W_1 \cap W_1}$ is handled similarly.

4. Proof that 8 relations are dyadically enumerable:
   (a) In dyadic notation, the successor of 1 is 2; the successor of 2 is 11; the successor of $x1$ is $x2$; and the successor of $x2$ is $y1$, where y is the successor of x. The relation "the successor of x is y" is represented by S in the elementary dyadic system whose initial formulas are the following:
   $S1, 2$
   $S2, 11$
   $Sx1, x2$
   $Sx, y \rightarrow Sx2, y1$
   (b) For easier readability, we write "$x < y$" instead of "$< x, y$". We take the following elementary dyadic system to represent this relation (note that the "S" here refers to the successor relation defined in (a)):
   $Sx, y \rightarrow x < y$
   $x < y \rightarrow y < z \rightarrow x < z$
   Thus "$<$" represents the relation "x is less than y".
   (c) Take the one scheme: $x = x$.
   (d) Take the system:
   $x < y \rightarrow x \leq y$
   $x = y \rightarrow x \leq y$
   (e) Take the system:
   $x < y \rightarrow x \neq y$
   $y < x \rightarrow x \neq y$
   (f) The relation $x + y = z$ is uniquely determined by the following conditions:
      (1) $x + 1 = x'$, where x' is the successor of x.
      (2) If y' is the successor of y, then $x + y' = (x + y)'$, where y' is the successor of y and $(x + y)'$ is the successor of $x + y$.
      Stated otherwise, if $x + y = z$, then $x + y' = z'$, where y' is the successor of y and z' is the successor of z. Thus the addition relation "$x + y = z$", which we will represent by "A" can be represented by taking the system of (a) for the successor relation, and adding the schemes:
      $Sx, y \rightarrow Ax, 1, y$
      $Ax, y, z \rightarrow Sy, u \rightarrow Sz, v \rightarrow Ax, u, v$
   (g) The relation $x \times y = z$ is uniquely determined by the following conditions:
      (1) $x \times 1 = x$
      (2) $x \times y' = (x \times y) + x$ (where y' is the successor of y)
      Thus, to represent the relation "$x \times y = z$" by, say, "M", we take the preceding system and adjoin the following:
      $Mx, 1, x$
      $Mx, y, z \rightarrow Az, x, w \rightarrow Sy, u \rightarrow Mx, u, w$

(h) The relation $x^y = z$ (among positive integers) is uniquely determined by the following conditions:
$x^1 = x$
$x^{y'} = x^y \times x$ (where y' is the successor of y)
We thus add the following to the last system, using "E" to represent the relation $x^y = z$:
$Ex, 1, x$
$Sy, u \to Ex, y, z \to Mx, z, w \to Ex, u, w$

5. I illustrate the proof of both claims with a proof for a relation $R(x, y)$ of two arguments. Suppose R is single-valued and dyadically enumerable. Let $\overline{R}(x, y)$ be the relation that $R(x, y)$ doesn't hold. We are to show that $\overline{R}$ is also dyadically enumerable. Well, since R is single-valued, to say that $R(x, y)$ doesn't hold is equivalent to saying that $R(x, z)$ holds for some z that is unequal to y (this is because one of the assumptions about R when we know that it is singe-valued, is that for every x it does hold for some value y; of course to be single-valued it can only hold for one such value). Thus we take a dyadic system in which the relation R is represented, say by "R", and where the inequality relation (which is dyadically enumerable, by (e) of problem 4) is also represented, say by "$\neq$", and we take a new predicate, say "$\overline{R}$", and add the axiom scheme $Rx, y \to z \neq y \to \overline{R}x, z$. Then "$\overline{R}$" represents the relation $\overline{R}(x, y)$.
6. All of them are dyadically solvable! The relations (a), (f), (g), and (h) are all single valued and dyadically enumerable, hence dyadically solvable by Problem 5. Also, by (c) and (e) of Problem 4, the relation $x = y$ and its complement $x \neq y$ are both dyadically enumerable, hence both are dyadically solvable. As for (b) and (d), the relations $x < y$ is the complement of the relation $y \leq x$, and since they are both dyadically enumerable, they are both dyadically solvable.
7. Proof: $x \in G_n$ iff $1Bx \wedge 2Ex \wedge {\sim}11Px \wedge {\sim}g_{n+1}Px$.
8. We first illustrate the proof for a special case: Let a, b, c be symbols of L and let a_0, b_0, c_0 be their dyadic Gödel numbers. Let X be the string $bx_1abx_2cx_1b$, where x_1 and x_2 are variables. Let $X*$ be the string $b_0x_1a_0b_0x_2c_0x_1b_0$ (which is the result of replacing each occurrence of a, b, c by their respective Gödel numerals.Then $x \in I_0(X)$ iff $(\exists x_1 \leq x)(\exists x_2 \leq x)(G_n(x_1) \wedge G_n(x_2) \wedge x = X*)$ i.e. iff $(\exists x_1 \leq x)(\exists x_2 \leq x)(G_n(x_1) \wedge G_n(x_2) \wedge x = b_0x_1a_0b_0x_2c_0x_1b_0)$. More generally, let X be any string compounded from symbols of L and variables $x_1, \ldots, x_t$. Let $X*$ be the result of replacing each symbol of L that is in X with its dyadic Gödel numeral. Then $x \in I_0(X)$ iff the following holds: $(\exists x_1 \leq x) \ldots (\exists x_t \leq x)(G_n(x_1) \wedge \cdots \wedge G_n(x_t) \wedge x = X*)$.
9. We recall that for each i, the set G_i is $\sum_0$.
   (a) $Seq(x)$ iff $G_m(x) \wedge g_mBx \wedge g_mEx \wedge {\sim}g_mg_mPx$
   (b) $x \in y$ iff $Seq(y) \wedge g_mxg_mPy \wedge G_n(x)$ (here m is one plus the number of symbols in L and n is the number of symbols in L).
   (c) $pr(x, y, z)$ iff $y \in z \wedge (\exists w \leq z)(wBz \wedge x \in w \wedge {\sim}yPw)$

10. Let b be the Gödel number of the implication sign. Then $Der(x, y, z)$ iff $y = xbz \wedge \sim bPx$.
11. Let $Y_1, \ldots, Y_r$ be the axiom schemes of (E), and for each $i \leq r$, let A_i be the set of $I_0(Y_i)$. By the substitution lemma, each of the sets $A_1, \ldots, A_r$ is $\sum_0$, hence so is their union $A_1 \cup \cdots \cup A_r$. Thus the set A of Gödel numbers of all the axioms of (E) is $\sum_0$.
    (a) $Pf(x)$ iff

$$Seq(x) \wedge (\forall y \leq x)[y \in x \supset (A(y) \vee$$
$$(\exists z \leq x)(\exists w \leq x)(pr(z, y, x) \wedge pr(w, y, x) \wedge Der(z, w, y))]$$

    (b) $ypfx$ iff $Pf(y) \wedge x \in y \wedge g_m x g_m E y$
    (c) $Pr(x)$ iff $\exists y(ypfx)$
    (d) Let H be a predicate that represents W in (E) and let h be the Gödel number of H. Then $x \in W_0$ iff $\exists y(ypfhx)$. [Note: $ypfhx$ can be written $(\exists z \leq y)(z = hx \wedge ypfz)$.]
12. If the set T of true sentences of first-order arithmetic were formally representable, then the set T_0 would be $\sum_1$, hence arithmetic, which it isn't, by Tarski's Theorem. Therefore, the set T is not formally representable.
13. Let us first note the following. Suppose $\mathcal{S}$ is $\sum_0$-complete and W is a formally representable set of expressions. Then the set W^* must be enumerable in $\mathcal{S}$, because W_0 is then $\sum_1$, hence W^* (which is $W_0^{\#}$) is also $\sum_1$ (by Problem 12, Chapter 11), and is thus the domain of a $\sum_0$-relation $R(x, y)$. Since the system is $\sum_0$-complete, the relation $R(x, y)$ is definable in the system, and therefore W^* is enumerable in the system.

    Now suppose $\mathcal{S}$ is a formal system that is $\sum_0$-complete. Let P be its set of provable sentences and let R be its set of refutable sentences. Since $\mathcal{S}$ is a formal system, the sets P and R are formally representable, hence the sets P^* and R^* are both enumerable in as seen above.

    (a) Thus there is a formula $A(x, y)$ that enumerates the set P^*. If the system is omega consistent, then by Theorem G_1 of Chapter 11, the sentence $\forall y(\sim A(\overline{p}, y))$ is undecidable, where p is the Gödel number of the formula $\forall y(\sim A(x, y))$. This proves Theorem G.
    (b) Suppose conditions L_1 and L_2 also hold. Then $\mathcal{S}$ satisfies the Rosser conditions, and so if the system is *simply consistent*, there is an undecidable sentence by Theorem R_1, Chapter 10.

13

Peano Arithmetic

In 1891, Giuseppe Peano published his famous postulates for the positive integers (which were then sometimes called "natural numbers," but nowadays the so-called "natural numbers" include zero). Here are the postulates (or axioms):

His undefined notions were 1 and the successor operation $S(n)$.

1. 1 is a natural number.
2. If n is a natural number, then so is $S(n)$.
3. If $S(n) = S(m)$, then $n = m$ (no two distinct natural numbers have the same successor).
4. $S(n) \neq 1$ (1 is not the successor of any natural number).
5. (Axiom of Mathematical Induction.) Suppose K is a set such that:
   (1) $1 \in K$.
   (2) For any natural number $n \in K$, the natural number $S(n)$ is also in K.

Then K contains all natural numbers.

Peano's so-called "axioms" do not constitute an axiom system in the modern sense of the term. They might aptly be called "informal axioms," which, like the so-called axioms of the ancient Greeks we see in Euclid's works, consisted of truths regarded as self-evident.

We now turn to the modern version of Peano Arithmetic in which the underlying logic is made explicit, and where axioms for addition and multiplication supplement those for the successor function. The axioms fall into three groups. Those of Group I are the axioms of *Propositional Logic*, which deals with the logical connectives $\sim$, $\wedge$, $\vee$, and $\supset$ (not, and, or, implies). Those of Group II are axioms of *First-Order Logic*, and are concerned with the quantifiers $\forall$ and $\exists$. The axioms of Group III are concerned with purely arithmetic notions – successors, plus and times. Equality and the relation of being less than (or of being less than or equal to in the case of our particular axiom system; but $<$, $>$, and $\geq$ can be easily defined from $\leq$, $=$, and the logical connective $\sim$).

The axiom system that we give here is rather similar to that given in Kleene [1952].

Concerning the axioms of Group II, for any variables x and y and any formula F, we say that x is bound by y in F if there is some formula G such that x has at least one occurrence in G and either $\forall yG$ or $\exists yG$ is part of F.

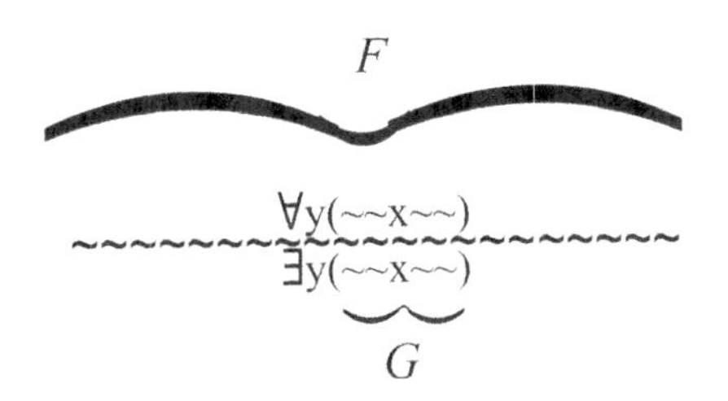

For any term t, variable x and formula F, we say that t is free for x in F if x is not bound in F by any variable that occurs in t. The purpose of these definitions is to be able to give a condition under which, if a term t with free variables in it is substituted for all free occurrences of a variable x in a formula G, then none of the variables that were free in t will be bound in the result of substituting t for free occurrences of x in G. We did not have to worry about this in First-Order Logic, because the only substitutions we did there were ones of parameters for free variables.

In the Group III Axioms, both here and later when we use them, we will put parentheses around atomic formulas when we feel it will make an axiom more easily understandable

Note: We use $X \equiv Y$ to abbreviate $(X \supset Y) \wedge (Y \supset X)$.

Axiom Schemes and Inference Rules of Peano Arithmetic:

Group I Axiom Schemes

A1. $(F \wedge G) \supset F$

A2. $(F \wedge G) \supset G$

A3. $[(F \wedge G) \supset H] \supset [F \supset (G \supset H)]$

A4. $[(F \supset G) \wedge (F \supset (G \supset H))] \supset (F \supset H)$

A5. $F \supset (F \vee G)$

A6. $G \supset (F \vee G)$

A7. $[(F \supset H) \wedge (G \supset H)] \supset ((F \vee G) \supset H)$

A8. $((F \supset G) \wedge (F \supset \sim G)) \supset \sim F$

A9. $\sim\sim F \supset F$

Group I Inference Rules

Modus Ponens $\qquad \dfrac{F, \quad F \supset G}{G}$

Group II Axiom Schemes

In these two schemes, t is a term free for x in F, and $F(t)$ is the result of substituting t for all free occurrences of x in $F(x)$.

A10. $\forall x F(x) \supset F(t)$
A11. $F(t) \supset \exists x F(x)$

Group II Inference Rules

In these two rules, C is a formula in which x has no free occurrences.

Rule 1. $$\frac{C \supset F(x)}{C \supset \forall x F(x)}$$

Rule 2. $$\frac{F(x) \supset C}{\exists x F(x) \supset C}$$

Group III Axiom Schemes

A12. $(x' = y') \supset (x = y)$
A13. $\sim(x' = 0)$
A14. $((x = y) \wedge (x = z)) \supset (y = z)$
A15. $(x = y) \supset (x' = y')$
A16. $x + 0 = x$
A17. $x + y' = (x + y)'$
A18. $x \times 0 = 0$
A19. $x \times y' = (x \times y) + x$
A20. $(F(0) \wedge \forall x(F(x) \supset F(x'))) \supset \forall x F(x)$ *Mathematical Induction*
A21. $(x \leq 0) \equiv (x = 0)$
A22. $(x \leq y') \equiv ((x \leq y) \vee (x = y'))$
A23. $(x \leq y) \vee (y \leq x)$
A24. $x = x$
A25. $F(\overline{n}) \supset ((x = \overline{n}) \supset F(x))$
A26. $\forall x {\sim} F(x) \supset {\sim}\exists x F(x)$

Until further notice, *provability* shall mean provability in Peano Arithmetic.

Proposition 1. Some basic facts about Peano Arithmetic:

(a) If $F(x)$ is provable, so is $\forall x F(x)$.
(b) If $F(x)$ is provable, so is $F(\overline{n})$ for any number n.
(c) If $F(x, y)$ is provable, so is $F(\overline{n}, \overline{m})$, for all n and m.

Problem 1. Prove Proposition 1.

We now wish to show that the set of provable formulas of Peano Arithmetic is formally representable, and hence that the set of corresponding Gödel numbers is $\sum_1$.

Problem 2. Successively show the following sets and relations to be formally representable. In fact, construct an elementary formal system (E) over the alphabet K of arithmetic in which they are all simultaneously represented.

(1) The set of strings of v's.
(2) The set of variables (of Peano Arithmetic).
(3) The relation "x and y are distinct variables".
(4) The set of numerals (Peano numerals).
(5) The set of terms.
(6) The set of atomic formulas.
(7) The set of formulas.
(8) The relation "t is a term, x is a variable, y is a term, and z is the result of substituting t for all occurrences of x in y".
(9) The relation "t is a term, x is a variable, f is a formula, and g is the result of substituting t for all *free* occurrences of x in f".
(10) The relation "x is a variable, f is a formula, and x does not occur *free* in f".
(11) The relation "x and y are variables, f is a formula, and x is not bound by y in f".
(12) The relation "t is a term, x is a variable, f is a formula, and t is free for x in f".
(13) The set of axioms.
(14) The set of provable formulas.

We have now shown that the set Pf of provable formulas of Peano Arithmetic is formally representable, and hence that the set Pf_0 of corresponding Gödel numbers is $\sum_1$, but we have not shown that the set P of provable *sentences* (formulas with no free variables) is formally representable, and hence that the set P_0 of Gödel numbers of the provable sentences is $\sum_1$. This can be done, but it is a bit complicated, and it is easier to show directly that P_0 is $\sum_1$, and this is what we shall do.

We temporarily let S be the set of all *sentences* of Peano Arithmetic, and we let S_0 be the corresponding set of Gödel numbers. The key to showing that P_0 is $\sum_1$ is to first show that S_0 is $\sum_1$.

Problem 3. Using the following three steps, prove that S_0, P_0 and R_0 are $\sum_1$.

(a) Prove that S_0 is $\sum_1$. [Hint: A formula X is a sentence iff no variable occurs free in it. It suffices that no variable that is part of X occurs free in X (since obviously no variable that is not part of X can occur free in X). Therefore it suffices that, for every variable z whose Gödel number is less than or equal to the Gödel number of X, the variable x has no free occurrence in X. This involves universal quantification only over the set of numbers less than or equal to the Gödel number of X.]

(b) Now prove that P_0 is $\sum_1$.
(c) Now show that R_0 is $\sum_1$, where R is the set of refutable sentences.

We have now shown that the set P_0 of Gödel numbers of the provable sentences of Peano Arithmetic is $\sum_1$, hence that P_0 is also arithmetic. But by Tarski's theorem, the set T_0 of Gödel numbers of the *true* sentences is not arithmetic. Therefore truth and provability don't coincide. This means that either some true sentence is not provable, or some provable sentence is not true. The latter alternative is ruled out by the reasonable assumption that the axioms of Peano Arithmetic are all true, and that the inference rules preserve truth, and hence that all provable sentences are true. Thus it follows that some true sentence X is not provable in Peano Arithmetic. Since X is true, its negation $\sim X$ is false, hence also not provable in Peano Arithmetic (again, under the assumption that Peano Arithmetic is a correct system), and so X is an undecidable sentence of Peano Arithmetic. We thus have:

Incompleteness Theorem for Peano Arithmetic, I, GT [Gödel, Tarski]. Peano Arithmetic, if correct, is incomplete. That is, if only true sentences are provable in Peano Arithmetic, then there is a true sentence X that is neither provable nor refutable in Peano Arithmetic.

Remarks. I am assuming that Peano Arithmetic is correct. We can actually *exhibit* a true but unprovable sentence of Peano Arithmetic as follows: We noted in the last chapter that if a set W of expressions is formally representable, then not only is W_0 a $\sum_1$ set, but given an elementary formal system (E) in which W is represented, we can actually exhibit a $\sum_1$ formula that expresses W_0. Now, we have given an elementary formal system (E) in which the set Pf of provable *formulas* of Peano Arithmetic is represented, hence we can find a $\sum_1$ formula that expresses Pf_0, and we have seen how, from this, to find a formula – call it "$P(x)$" – that expresses the set P_0 of Gödel numbers of the provable *sentences* of Peano Arithmetic. Since the formula $P(x)$ is arithmetic, so is its negation $\sim P(x)$, which expresses the complement $\widetilde{P_0}$ of P_0. Then, as explained in Chapter 11, we know how to find a Gödel sentence X for $\widetilde{P_0}$ (a sentence that is true iff it is not provable). Thus X is true iff X is not provable, and this is the sentence we seek. Of course I could exhibit X explicitly, but his would be quite tedious, and it seems to me, pointless.

Next, I want to turn to the Gödel incompleteness proof for Peano Arithmetic, which replaces the hypothesis of "correctness" by the weaker hypothesis of omega consistency, and then look at the Rosser proof, which uses the still weaker hypothesis of simple consistency.

The reader may wonder why we should do this, since we have already shown Peano Arithmetic to be incomplete. Well, the incompleteness proof we have so far given for Peano Arithmetic is not what is termed a "finitary" proof. What is a finitary proof? No uniform definition has yet been given, but all the proposed definitions agree that any proof that assumes the existence of an infinite set is not a finitary proof. The notion of *truth*, for sentences of first-order arithmetic, is not a finitary notion. Although it is well-defined, it is not generally subject to verification, since it involves quantification over infinitely many natural numbers. To check whether a universal sentence $\forall x F(x)$ is true, generally involves checking the infinitely many sentences $F(\overline{0}), F(\overline{1}), \ldots, F(\overline{n}), \ldots$, which cannot be done in a finite length of time. By contrast, the notion of a sequence of formulas constituting a proof in a formal system is a finitary one.

Thus, there are some mathematicians who would not accept the incompleteness proof so far given, but would probably accept the Gödel proof, and certainly accept the Rosser proof, which is unquestionably finitary. For the sake of the record, let me state that I myself do not belong to the group that rejects non-finitary methods. I totally accept the incompleteness proof already given, which involves the notion of truth.

Nevertheless, I find the Gödel and Rosser proofs of equal interest for different reasons. Both proofs are extremely ingenious, and the Gödel proof is necessary as a prerequisite to Gödel's Second Incompleteness Theorem, which I will discuss briefly later, and which, roughly stated, says that Peano Arithmetic, if consistent, cannot prove its own consistency.

Now for the Gödel proof that if Peano Arithmetic is omega consistent, it is incomplete. We have already shown that Peano Arithmetic is a formal system, hence by Theorem G of Chapter 12, all that remains is to show that all $\sum_0$ relations are definable in Peano Arithmetic. To this end, it suffices to show that all true $\sum_0$ sentences are provable in Peano Arithmetic. [The notion of *truth* for $\sum_0$ sentences is quite finitary, since it involves only finite quantifications. One can always check whether a $\sum_0$ sentence is true or not.] We shall say that a $\sum_0$ sentence is *correctly decidable* (in Peano Arithmetic) is it is either true and provable in Peano Arithmetic or false and refutable in Peano Arithmetic. We shall show that all $\sum_0$ sentences are correctly decidable in Peano Arithmetic, and hence that all true $\sum_0$ sentences are definable in Peano Arithmetic. We will in fact establish a more general result that we will need for a later chapter.

The following formula schemes play an important role in much of modern research:

$\Omega 1.$ $\overline{m} + \overline{n} = \overline{k}$, where $m + n = k$
$\Omega 2.$ $\overline{m} \times \overline{n} = \overline{k}$, where $m \times n = k$
$\Omega 3.$ $\sim(\overline{m} = \overline{n})$, where $m \neq n$
$\Omega 4.$ $x \leq \overline{n} \equiv (x = \overline{0} \vee \cdots \vee x = \overline{n})$
$\Omega 5.$ $x \leq \overline{n} \vee \overline{n} \leq x$

Note: We will often abbreviate $\sim(x = y)$ as $x \neq y$.

Lemma R. $\Omega 1$ through $\Omega 5$ are all provable in Peano Arithmetic without A20.[2]

[2] Though we use mathematical induction in reasoning *about* the system!

Problem 4. Prove Lemma R.

The schemes Ω1 through Ω5 are the arithmetic axioms of a system of Raphael Robinson [1950] known as System (R). A system $\mathcal{S}$ is called an *extension* of (R) if Ω1 through Ω5 are all provable in $\mathcal{S}$. We have just shown that Peano Arithmetic is an extension of (R) (by Lemma R). We now wish to show that not only is Peano Arithmetic $\sum_0$ complete, but that every extension of (R) is $\sum_0$ complete.

Problem 5. Prove that every true *atomic* $\sum_0$ sentence is provable in any extension of (R) (hence provable in Peano Arithmetic).

We recall that we are calling a $\sum_0$ sentence *correctly decidable* in a system if it is either true and provable in the system, or false and refutable in the system.

Problem 6. Now prove that every *atomic* $\sum_0$ sentence is correctly decidable in any extension of (R) (hence correctly decidable in Peano Arithmetic).

Problem 7. Show that for an extension $\mathcal{S}$ of (R), if $F(\overline{0}), F(\overline{1}), \ldots, F(\overline{n})$ are all provable in $\mathcal{S}$, so is $(\forall x \leq \overline{n})F(x)$.

Problem 8. Now prove that every extension $\mathcal{S}$ of (R) is $\sum_0$ complete.

We have now established most of the results necessary for the proofs of the Gödel and Rosser incompleteness theorems for Peano Arithmetic. We recall Theorem G of Chapter 12, which is that any formal $\sum_0$ complete system which is omega consistent has an undecidable sentence. Well, we have proved that Peano Arithmetic is a formal system and that it is $\sum_0$ complete, and so we have:

Incompleteness Theorem for Peano Arithmetic, II, G [Gödel]. Peano Arithmetic, if omega consistent, is incomplete.

This theorem is known as Gödel's First Incompleteness Theorem, but is often just referred to as "Gödel's Theorem."

Now for the Rosser Incompleteness proof for Peano Arithmetic. By Theorem *R* of Chapter 12, to show that Peano Arithmetic, if simply consistent, is incomplete, it suffices to show that Peano Arithmetic is formal and $\sum_0$ complete, which we have done, and that for any formula $F(x)$ and any number n:

L_1: If $F(\overline{0}), F(\overline{1}), \ldots, F(\overline{n})$ are all provable in Peano Arithmetic, so is $(\forall x \leq \overline{n})F(x)$.

L_2: $\forall x(x \leq \overline{n} \vee \overline{n} \leq x)$ is provable in Peano Arithmetic.

Problem 9. Prove L_1 and L_2 above.

We have now proved:

Incompleteness Theorem for Peano Arithmetic, III, R [Rosser]. Peano Arithmetic, if simply consistent, is incomplete.

Discussion

I would like to conclude this chapter by saying that I regard the incompleteness of Peano Arithmetic to be of less significance than the fact that the set of *true* sentences is not only not solvable, but not even formally representable. This means that not only is there no purely mechanical procedure to decide which sentences are true and which are not, but there is even no mechanical procedure which will generate the set of all and only those sentences that are true! It is really the *true* sentences that interest the working mathematician, not the ones provable in Peano Arithmetic, If a sentence is provable or refutable in Peano Arithmetic, all well and fine, since it would answer the question of whether it is true or not. But if the sentence is neither provable nor refutable in Peano Arithmetic, nothing can to gleaned as to its truth.

Despite the incompleteness of Peano Arithmetic, the system has many other interesting features, which we will explore later on.

Solutions to the Problems of Chapter 13

1. It is assumed that $F(x)$ is provable in (a) and (b).
   (a) Take any closed formula (sentence) C that is provable. Since $F(x)$ is provable, so is $C \supset F(x)$ by Propositional Logic (this actually follows from the T_5 that has been proved; but anything propositionally true is provable here, since we have all the axioms and inference rules of the system of Propositional Logic, which was proved complete in Chapter 7). Hence $C \supset \forall x F(x)$, by Rule 1, Group II. Since C is also provable, so is $\forall x F(x)$ by Modus Ponens.
   (b) Since $F(x)$ is provable, so is $\forall x F(x)$ by (a). Also, $\forall x F(x) \supset F(\overline{n})$ is provable by Axiom A10, since $\overline{n}$ is obviously free for x in $F(x)$. Since the formulas $\forall x F(x) \supset F(\overline{n})$ and $\forall x F(x)$ are both provable, so is $F(\overline{n})$ by Modus Ponens.
   (c) Suppose that $F(x, y)$ is provable. Then so is $F(\overline{n}, y)$ by (b). Hence so is $F(\overline{n}, \overline{m})$, again by (b).
   Note: Indeed, if $F(x_1, \ldots, x_k)$ is provable, so is $F(\overline{n_1}, \ldots, \overline{n_k})$, which can be shown by mathematical induction.
2. In the elementary formal system (E) which follows, in which we represent the various sets and relations, we introduce the predicates and axiom schemes in groups, first explaining what each newly introduced predicate is to represent. We must not confuse variables of (E) with variables of Peano Arithmetic. For the former, we use the letters x, y, z, w, t, f, g, h, with or without subscripts. Also,

"$\supset$" is the implication sign of Peano Arithmetic, whereas "$\rightarrow$" is the implication sign of the elementary formal system (E).

(1) st will represent the set of strings of v's.

$st\ v$

$st\ x \rightarrow st\ xv$

(2) V will represent the set of variables (of Peano Arithmetic).

$st\ x \rightarrow V(x)$

(3) Dv will represent the relation "x and y are distinct variables".

$st\ x \rightarrow st\ y \rightarrow Dv\ x, xy$

$Dv\ x, y\ \rightarrow Dv\ y, x$

(4) N will represent the set of numerals.

$N\ 0$

$N\ x\ \rightarrow N\ x'$

(5) Tm will represent the set of terms.

$V\ x \rightarrow Tm\ x$

$N\ x \rightarrow Tm\ x$

$Tm\ x \rightarrow Tm\ y \rightarrow Tm\ (x + y)$

$Tm\ x \rightarrow Tm\ y \rightarrow Tm\ (x \times y)$

$Tm\ x \rightarrow Tm\ x'$

(6) F_0 will represent the set of atomic formulas.

$Tm\ x \rightarrow Tm\ y \rightarrow F_0\ x = y$

$Tm\ x \rightarrow Tm\ y \rightarrow F_0\ x \leq y$

(7) F will represent the set of formulas.

$F_0\ x \rightarrow F\ x$

$F\ x \rightarrow F{\sim}x$

$F\ x \rightarrow F\ y \rightarrow F\ (x \wedge y)$

$F\ x \rightarrow F\ y \rightarrow F\ (x \vee y)$

$F\ x \rightarrow F\ y \rightarrow F\ (x \supset y)$

$V\ x \rightarrow F\ y \rightarrow F\ \forall xy$

$V\ x \rightarrow F\ y \rightarrow F\ \exists xy$

(8) S_0 will represent the relation "t is a term, x is a variable, y is a term and z is the result of substituting t for all occurrences of x in y".

$Tm\ t \rightarrow V\ x \rightarrow S_0\ t, x, x, t$

$Tm\ t \rightarrow V\ x \rightarrow Ny \rightarrow S_0\ t, x, y, y$

$Tm\ t \rightarrow D\ x, y \rightarrow S_0\ t, x, y, y$

$S_0\ t, x, y, z \rightarrow S_0\ t, x, y', z'$

$S_0\ t, x, y, z \rightarrow S_0\ t, x, y_1, z_1 \rightarrow S_0\ t, x, y + y_1, z + z_1$

$S_0\ t, x, y, z \rightarrow S_0\ t, x, y_1, z_1 \rightarrow S_0\ t, x, y \times y_1, z \times z_1$

(9) S will represent the relation "t is a term, x is a variable, f is a formula and g is the result of substituting t for all free occurrences of x in f".

$S_0\ t, x, y, z \rightarrow S_0\ t, x, y_1, z_1 \rightarrow S\ t, x, y = y_1, z = z_1$

$S_0\ t, x, y, z \rightarrow S_0\ t, x, y_1, z_1 \rightarrow S\ t, x, y \leq y_1, z \leq z_1$

$S\, t, x, f, g \to S\, t, x, {\sim}f, {\sim}g$
$S\, t, x, f, g \to S\, t, x, f_1, g_1 \to S\, t, x, (f \wedge f_1), (g \wedge g_1)$
$S\, t, x, f, g \to S\, t, x, f_1, g_1 \to S\, t, x, (f \vee f_1), (g \vee g_1)$
$S\, t, x, f, g \to S\, t, x, f_1, g_1 \to S\, t, x, (f \supset f_1), (g \supset g_1)$
$S\, t, x, f, g \to D\, x, y \to S\, t, x, \forall y f, \forall y g$
$S\, t, x, f, g \to D\, x, y \to S\, t, x, \exists y f, \exists y g$
$Tm\, t \to F\, f \to V\, x \to S\, t, x, \forall x f, \forall x f$
$Tm\, t \to F\, f \to V\, x \to S\, t, x, \exists x f, \forall \exists f$

(10) *Noc* will represent the relation "x is a variable, f is a formula and x does not occur free in f".
$D\, y, x \to S\, y, x, f, f \to Noc\, x, f$

(11) $\overline{B}$ will represent the relation "x and y are variables, f is a formula and x is not bound by y in f".
$V\, x \to V\, y \to Tm\, z \to Tm\, w \to \overline{B}\, x, y, z = w$
$V\, x \to V\, y \to Tm\, z \to Tm\, w \to \overline{B}\, x, y, z \leq w$
$\overline{B}\, x, y, f \to \overline{B}\, x, y, {\sim}f$
$\overline{B}\, x, y, f \to \overline{B}\, x, y, (f \wedge g)$
$\overline{B}\, x, y, f \to \overline{B}\, x, y, (f \vee g)$
$\overline{B}\, x, y, f \to \overline{B}\, x, y, (f \supset g)$
$\overline{B}\, x, y, f \to D\, y, z \to \overline{B}\, x, y, \forall z f$
$\overline{B}\, x, y, f \to D\, y, z \to \overline{B}\, x, y, \exists z f$
$\overline{B}\, x, y, f \to Noc\, x, f \to \overline{B}\, x, y, \forall y f$
$\overline{B}\, x, y, f \to Noc\, x, f \to \overline{B}\, x, y, \exists y f$

(12) E will represent the relation "t is a term, x is a variable, f is a formula and t is free for x in f".
$\overline{B}\, x, y, f \to E\, y, x, f$
$V\, x \to N\, y \to F\, f \to E\, y, x, f$
$E\, t, x, f \to E\, t', x, f$
$E\, t, x, f \to E\, t_1, x, f \to E\, (t + t_1), x, f$
$E\, t, x, f \to E\, t_1, x, f \to E\, (t \times t_1), x, f$

(13) A will represent the set of Axioms [Note: for the nine axioms of Propositional Logic (Group I), I give the solution only to the first, since the other eight axioms are obviously similar, and are left to the reader.]
1 $F\, f \to F\, g \to A\, (f \wedge g) \supset f$
2 $F\, f \ldots$
⋮
10 $E\, t, x, f \to S\, t, x, f, g \to A\, (\forall x f) \supset g$
11 $E\, t, x, f \to S\, t, x, f, g \to A\, g \supset (\exists x f)$
12 $Tm\, x \to Tm\, y \to A\, (x' = y') \supset (x = y)$
13 $Tm\, x \to A\ {\sim}(x' = 0)$
14 $Tm\, x \to Tm\, y \to Tm\, z \to A\, ((x = y) \wedge (x = z)) \supset (y = z)$
15 $Tm\, x \to Tm\, y \to A\, (x = y) \supset (x' = y')$

16 $Tm\ x \to A\ \ x+0=x$
17 $Tm\ x \to Tm\ y \to A\ x+y'=(x+y)'$
18 $Tm\ x \to A\ \ x\times 0=0$
19 $Tm\ x \to Tm\ y \to A\ x\times y'=(x\times y)+x$
20 $V\ x \to S\,0, x, f, g \to S\,x', x, f, h \to A\,(g \wedge \forall x(f \supset h)) \supset \forall xf$
21 $Tm\ x \to A\ \ (x\leq 0)\ \equiv\ (x=0)$
22 $Tm\ x \to Tm\ y \to A\ (x\leq y')\ \equiv\ ((x\leq y)\vee(x=y))$
23 $Tm\ x \to Tm\ y \to A\ (x\leq y)\vee(y\leq x)$
24 $Tm\ x \to A\ \ x=x$
25 $N\ y \to S\ y, x, f, g \to A\ g\supset(x=y\ \supset\ f)$
26 $F\ f \to V\ x \to A\ \forall x{\sim}f \supset {\sim}\exists xf$

(14) P will represent the set of provable formulas.
$A\ f \to P\ f$
$P\ f \to P\ (f\supset g) \to P\ g$
$Noc\ x, c \to F\ f \to P\ (c\ \supset f) \to P\ (c\supset \forall xf)$
$Noc\ x, c \to F\ f \to P\ (f\ \ \supset c) \to P\ (\exists xf\ \supset c)$

3. Let F be the set of formulas and let F_0 be the corresponding set of Gödel numbers. Since F is formally representable, as we have seen, then F_0 is $\sum_1$ (by Theorem A of Chapter 12).

Let V be the set of variables of Peano Arithmetic. Since V is formally representable, V_0 is $\sum_1$.

We recall the formally representable relation $Noc(x, y)$ ("x is a formula and y is a variable that does not occur free in x"). Let $NocG(x, y)$ be the corresponding relation between Gödel numbers (i.e. $NocG(x, y)$ means that x and y are Gödel numbers of a formula u and a variable v, respectively, such that $Noc(u, v)$ holds). Thus $NocG(x, y)$ is a $\sum_1$ relation.

(a) We are letting S_0 be the set of Gödel numbers of the sentences, and we are to show that S_0 is $\sum_1$. Well, $x \in S_0$ iff the following condition holds:

$$F_0(x)\wedge(\forall y\leq x)(V_0(y)\supset NocG(x, y)).$$

We leave it to the reader to show that the above condition is $\sum_1$ (recall Problem 10 of Chapter 11).

(b) The set Pf of provable *formulas* has been shown to be formally representable, so that the set Pf_0 of corresponding Gödel numbers is $\sum_1$. Then

$$x\in P_0 \quad \text{iff} \quad Pf_0(x)\wedge S_0(x).$$

(c) $x \in R_0$ iff $(7{*}z)\in P_0$ (7 is the Gödel number of "$\sim$".)

4. We first note that for any number n, the numeral $\overline{n}'$ is the same expression as $\overline{n+1}$. Now to prove the formulas Ω1 through Ω5.

Ω1. The formula $x+y'=(x+y)'$ is provable (it is Axiom A17), hence, by Proposition 1(c), for any numbers m and n, the sentence $\overline{m}+\overline{n}'=(\overline{m}+\overline{n})'$

is provable. Therefore, for any number q, if $\overline{m}+\overline{n}=\overline{q}$ is provable, so is $\overline{m}+\overline{n}'=\overline{q}'$ and hence so is $\overline{m}+\overline{n+1}=\overline{q+1}$. Thus

(1) If $\overline{m}+\overline{n}=\overline{q}$ is provable, so is $\overline{m}+\overline{n+1}=\overline{q+1}$.

Now, $\overline{m}+\overline{0}=\overline{m}$ is provable (by A16 and Proposition 1(b)), and so by (1) we can successively prove $\overline{m}+\overline{1}=\overline{m+1}, \overline{m}+\overline{2}=\overline{m+2}, \ldots \overline{m}+\overline{n}=\overline{m+n}$. [We have informally used mathematical induction.]

Ω2. The proof is similar here, using A19 and A18 instead of A17 and A16.

Ω3. We are to show that $\sim(\overline{m}=\overline{n})$ is provable in Peano Arithmetic, whenever $m \neq n$. We will do this for the case of m and n such that $m > n$. Define k to be the number $m-n$. Clearly k is positive and $m=k+n$. Thus we need to show that the sentence $\overline{k+n} \neq \overline{n}$ is provable when $m \neq n$ is true.

From the axiom $(x'=y') \supset (x=y)$ [Axiom A12] and Proposition 1(c), it follows that for any numbers m and n, the sentence $\overline{m}'=\overline{n}' \supset \overline{m}=\overline{n}$ is provable, hence by Propositional Logic, so is $\overline{m} \neq \overline{n} \supset \overline{m}' \neq \overline{n}'$, and thus also $\overline{m} \neq \overline{n} \supset \overline{m+1} \neq \overline{n+1}$. Hence

(1) If $\overline{m} \neq \overline{n}$ is provable, so is $\overline{m+1} \neq \overline{n+1}$.

Next, from the axiom $x' \neq 0$ [Axiom A13] and Proposition 1(b), it follows that for any number j, the sentence $\overline{j}' \neq 0$ is provable, and hence that for our *positive* number k defined above, the sentence $\overline{k} \neq 0$ is provable (since $\overline{k}$ is $\overline{k-1}'$). Then by (1) we can successively prove $\overline{k+1} \neq \overline{1}$, $\overline{k+2} \neq \overline{2}, \ldots, \overline{k+n} \neq \overline{n}$, which is all we needed (because $m=k+n$).

Ω4. We are to show that the formula $x \leq \overline{n} \equiv (x=\overline{0} \vee \cdots \vee x=\overline{n})$ is provable in Peano Arithmetic. We will do this by mathematical induction.

For $n=0$, the formula $x \leq \overline{0} \equiv x=\overline{0}$ is an axiom [Axiom A21], hence is provable. Now suppose n is such that

(1) $x \leq \overline{n} \equiv (x=\overline{0} \vee \cdots \vee x=\overline{n})$ is provable.

To show the second induction premise is true, we must show this implies that $x \leq \overline{n+1} \equiv (x=\overline{0} \vee \cdots \vee x=\overline{n+1})$ is provable.

Now, $x \leq y' \equiv (x \leq y \vee x=y')$ is an axiom [Axiom A22], hence provable. Then, by Proposition 1(b), $x \leq \overline{n}' \equiv (x \leq \overline{n} \vee x=\overline{n}')$ is provable. It follows that

(2) $x \leq \overline{n+1} \equiv (x \leq \overline{n} \vee x=\overline{n+1})$ is provable.

By (1) and (2) it follows by Propositional Logic that

$$x \leq \overline{n+1} \equiv (x=\overline{0} \vee \cdots \vee x=\overline{n} \vee x=\overline{n+1}),$$

which completes the induction and proves our case.

$\Omega 5$. $x \leq \overline{n} \vee \overline{n} \leq x$ is immediate from Proposition 1(b) and the axiom $x \leq y \vee y \leq x$ [Axiom A23].

5. We are to show that every true atomic $\sum_0$ sentence is provable in Peano Arithmetic. It must first be shown that for any constant term t, the sentence $t = \overline{n}$ is provable in Peano Arithmetic, where n is the natural number designated by t. This is easily established by mathematical induction on the number of occurrences of the mathematical operations (plus, times and successor) in the term t.

   Now that we know that for *any* constant term t designating the natural number n, the formula $t = \overline{n}$ is provable, we will show that if m_1 and m_2 are the natural numbers designated by the constant terms t_1 and t_2 respectively, then if $t_1 = t_2$ is true – which means that t_1 and t_2 designate the same natural number – then $t_1 = t_2$ is provable in Peano Arithmetic (and afterwards we must also show the same result for the similar but slightly more complicated case of the formula $t_1 \leq t_2$).

   Now if $t_1 = t_2$ is true and both t_1 and t_2 designate the natural number n, by the above we have that $t_1 = \overline{n}$ and $t_2 = \overline{n}$ are both provable. So, by the substitution of equals (derivable from Axiom A14), we have that $t_1 = t_2$ is provable. Now for the slightly more complicated case. Assume $t_1 \leq t_2$ is true, where t_1 designates m_1 and t_2 designates m_2. This means that $m_1 \leq m_2$ (by the definition of truth for systems of Elementary Arithmetic). We also know that $t_1 = \overline{m_1}$ and $t_2 = \overline{m_2}$ are both provable. So (by the substitution of equals), if we can show that $t_1 \leq \overline{m_2}$ is provable, we will have the result that we need. But by Ω_4 we know that

$$t_1 \leq \overline{m_2} \equiv (t_1 = \overline{0} \vee t_1 = \overline{1} \vee \cdots \vee t_1 = \overline{m_1} \vee \cdots \vee t_1 = \overline{m_2}).$$

   But the formula on the right side of the equivalence is clearly provable, since it contains the subformula $t_1 = \overline{m_1}$ that is provable (by assumption). So, since we have the full proof apparatus of Propositional Logic in Peano Arithmetic, $t_1 \leq \overline{m_2}$ must be provable, and consequently, as we remarked earlier, $t_1 \leq t_2$ is provable as well.

6. Suppose the system $\mathcal{S}$ is an extension of (R). We have shown that all true atomic sentences are provable in $\mathcal{S}$. It remains to show that all false atomic $\sum_0$ sentences are refutable in $\mathcal{S}$.

   (1) Consider a false $\sum_0$ sentence of the form $\overline{m} + \overline{n} = \overline{q}$. Since it is false, $m + n \neq q$. Hence $\overline{m+n} \neq \overline{q}$ is provable in $\mathcal{S}$ by $\Omega 3$. Also, $\overline{m} + \overline{n} = \overline{m+n}$ is provable by $\Omega 1$. Since also $\sim(\overline{m+n} = \overline{q})$ is provable [its abbreviation, $\overline{m+n} \neq \overline{q}$, was shown to be provable], then, by Propositional Logic, so is $\sim(\overline{m} + \overline{n} = \overline{q})$. Thus $\overline{m} + \overline{n} = \overline{q}$ is refutable.

   (2) The proof that any false sentence of the form $\overline{m} \times \overline{n} = \overline{q}$ is refutable is similar, using $\Omega 2$ instead of $\Omega 1$.

   (3) Suppose $\overline{m} = \overline{n}$ is false. Then $m \neq n$, hence $\overline{m} \neq \overline{n}$ is provable (by $\Omega 3$), and so $\overline{m} = \overline{n}$ is refutable.

(4) Consider a false $\sum_0$ sentence of the form $\overline{n}' = \overline{m}$. Since it is false, then $n+1 \neq m$, and so by $\Omega 3$ then sentence $\overline{n+1} \neq \overline{m}$ is provable, but this sentence is $\overline{n}' \neq \overline{m}$. Thus $\overline{n}' = \overline{m}$ is refutable.

(5) Finally, consider a false $\sum_0$ sentence of the form $\overline{m} \leq \overline{n}$. Since it is false, then all the sentences $\overline{m} = \overline{0}$, $\overline{m} = \overline{1}, \ldots \overline{m} = \overline{n}$ are all false, hence they are all refutable (by $\Omega 3$). Therefore, by Propositional Logic, the sentence $\overline{m} = \overline{0} \vee \cdots \vee \overline{m} = \overline{m} \vee \cdots \vee \overline{m} = \overline{n}$ is refutable. Also, the sentence $\overline{m} \leq \overline{n} \equiv (\overline{m} = \overline{0} \vee \cdots \vee \overline{m} = \overline{n})$ is provable (by $\Omega 4$). Hence by Propositional Logic $\overline{m} \leq \overline{n}$ is refutable in $\mathcal{S}$.

7. Suppose the system $\mathcal{S}$ is an extension of (R). By $\Omega 4$

$$(1) \quad x \leq \overline{n} \equiv (x = \overline{0} \vee \cdots \vee x = \overline{n})$$

is provable in $\mathcal{S}$. Also, by Axiom A25, the formula $F(\overline{m}) \supset (x = \overline{m} \supset F(x))$ is provable for any number m. Hence if $F(\overline{m})$ is provable, so is $x = \overline{m} \supset F(x)$.

Now, suppose if $F(\overline{0}), F(\overline{1}), \ldots, F(\overline{n})$ are all provable in $\mathcal{S}$. Then by Propositional Logic, the following is provable:

$$(2) \quad (x = \overline{0} \vee \cdots \vee x = \overline{n}) \supset F(x)$$

From (1) and (2) it follows that $x \leq \overline{n} \supset F(x)$ is provable, hence, by Proposition 1(a), so is $\forall x(x \leq \overline{n} \supset F(x))$, and this is the sentence $(\forall x \leq \overline{n})F(x)$.

8. We prove this by complete induction on the degree (number of occurrences of logical connectives and quantifiers) of the $\sum_0$ sentence.

(a) Any $\sum_0$ sentence of degree zero is an atomic $\sum_0$ sentence, hence it is correctly decidable by Problem 6.

(b) Now suppose that d is a positive number such that all $\sum_0$ sentences of degree less than d are correctively decidable.

Let X be a $\sum_0$ sentence of degree d. It must be of one of the forms $\sim Y$, $Y \wedge Z, Y \vee Z, Y \supset Z, (\forall x \leq \overline{n})F(x), (\exists x \leq \overline{n})F(x)$, where $Y, Z, F(x)$ are of lower degree than d.

Suppose X is one of the forms $\sim Y, Y \wedge Z, Y \vee Z, Y \supset Z$. By Propositional Logic, it is obvious that if Y is correctly decidable, so is $\sim Y$, and if Y and Z are both correctly decidable, so are $Y \wedge Z, Y \vee Z$, and $Y \supset Z$. Since Y and Z are of degree less than d, they are both correctly decidable, by the induction hypothesis. Thus if X is of any of the forms $\sim Y, Y \wedge Z, Y \vee Z, Y \supset Z$, then X is correctly decidable.

Now let us consider the case that X is of the form $(\forall x \leq \overline{n})F(x)$. Then $F(x)$ is of degree less than d, and for any number m, the sentence $F(\overline{m})$ is of lower degree than d, and hence by the induction hypothesis is correctly decidable. Suppose $(\forall x \leq \overline{n})F(x)$ is true. Then all the sentences $F(\overline{0}), F(\overline{1}), \ldots, F(\overline{n})$ are true, hence correctly decidable, hence provable (since they are true) and therefore $(\forall x \leq \overline{n})F(x)$ is provable (by Problem 7).

Now consider the case that the formula $(\forall x \leq \overline{n})F(x)$ is false. Then for some $m \leq n$, the sentence $F(\overline{m})$ is false, and hence is refutable. Since $m \leq n$, the $\sum_0$ sentence $\overline{m} \leq \overline{n}$ is true, and thus provable. Since $\overline{m} \leq \overline{n}$ and $\sim F(\overline{m})$ are both provable, then

(1) $\overline{m} \leq \overline{n} \supset F(\overline{m})$ is refutable (by Propositional Logic). Also,
(2) $\forall x(x \leq \overline{n} \supset F(x)) \supset (\overline{m} \leq \overline{n} \supset F(\overline{m}))$ is provable [Axiom A10].

From (1) and (2) it follows that $\forall x(x \leq \overline{n} \supset F(x))$ is refutable, and this sentence is $(\forall x \leq \overline{n})F(x)$.

Lastly, let us consider the case that X is of the form $(\exists x \leq \overline{n})F(x)$. Suppose it is true. Then for some $m \leq n$, the sentence $F(\overline{m})$ are true, hence provable (since it is of degree $\leq d$). Also, $\overline{m} \leq \overline{n}$ is provable, and so $\overline{m} \leq \overline{n} \wedge F(\overline{m})$ is provable, as is $\overline{m} \leq \overline{n} \supset F(\overline{m})$ [by Prepositional Logic]. Also, $(\overline{m} \leq \overline{n} \supset F(\overline{m})) \supset \exists x(x \leq \overline{n} \wedge F(x))$ is provable [by Axiom A11]. And so, by Propositional Logic, $\exists x(x \leq \overline{n} \wedge F(x))$ is provable. Thus $(\exists x \leq \overline{n})F(x)$ is provable.

Now suppose that $(\exists x \leq \overline{n})F(x)$ is false. Then for every $m \leq n$, the sentence $F(\overline{m})$ is false, hence refutable. Thus all the sentences $F(\overline{0}), F(\overline{1}), \ldots, F(\overline{n})$ are refutable. Since $\sim F(\overline{0}), \sim F(\overline{1}), \ldots, \sim F(\overline{n})$ are all provable, so is $(\forall x \leq \overline{n}) \sim F(x)$ (by Problem 7), and this is the sentence $\forall x(x \leq \overline{n} \supset \sim F(x))$. Therefore the open formula $x \leq \overline{n} \wedge \sim F(x)$ is provable (using Axiom A10), hence by Propositional Logic, so is $\sim(x \leq \overline{n} \wedge F(x))$. Then, by Proposition 1(a), $\forall x \sim(x \leq \overline{n} \wedge F(x))$ is provable. Also,

$$\forall x \sim(x \leq \overline{n} \wedge F(x)) \supset \sim\exists x(x \leq \overline{n} \wedge F(x))$$

is provable (by A26), and so $\sim\exists x(x \leq \overline{n} \wedge F(x))$ is provable, and so is the sentence $\sim(\exists x \leq \overline{n})F(x)$. Thus $(\exists x \leq \overline{n})F(x)$ is refutable.

This completes the induction.

9. Proofs of the two conditions:
   (a) Suppose $F(\overline{0}), F(\overline{1}), \ldots, F(\overline{n})$ are all provable. Then from Axiom A25 it follows by modus ponens that $x = \overline{0} \supset F(x), \ldots, x = \overline{n} \supset F(x)$ are all provable. Then, by Propositional Logic, we have:
   (1) $(x = \overline{0} \vee \cdots \vee x = \overline{n}) \supset F(x)$ is provable. Next, from $\Omega 4$ of Lemma R, we have:
   (2) $x \leq \overline{n} \supset (x = \overline{0} \vee \cdots \vee x = \overline{n})$ is provable.
   By Propositional Logic, from (1) and (2) we get
   (3) $x \leq \overline{n} \supset F(x)$ is provable.
   Then by Proposition 1(a) it follows that $\forall x(x \leq \overline{n} \supset F(x))$ is provable, and this sentence is $(\forall x \leq \overline{n})F(x)$.
   (b) By $\Omega 5$ the formula $x \leq \overline{n} \vee \overline{n} \leq x$ is provable. Hence $\forall x(x \leq \overline{n} \vee \overline{n} \leq x)$ is provable by Proposition 1(a).

14

Further Topics

Gödel's so-called *Second Incompleteness Theorem*, which is almost as renowned as the incompleteness theorem of his we have been studying, roughly stated, says that for Peano Arithmetic and related systems, if they are (simply) consistent, they cannot prove their own consistency. A more precise formulation of this theorem will be given in Part II of this chapter. First we must consider some preliminaries.

Part I – Diagonalization and Fixed Points

Recursive Relations

There are many different definitions in the literature of a number set or relation being *recursive*, but they are all equivalent – in fact all are equivalent to the relation and its complement being $\sum_1$. Some treatments *define* a relation to be *recursively enumerable* if it is $\sum_1$ and this is the course we will take. Thus we shall henceforth use the terms "$\sum_1$" and "recursively enumerable" synonymously. A set or relation is called *recursive* if it and its complement are both recursively enumerable, i.e. if both are $\sum_1$.

Proposition 1. If $\mathcal{S}$ is an extension of (R), then all recursive sets and relations are definable in $\mathcal{S}$.

Problem 1. Prove Proposition 1.

Strong Definability of Functions

A formula $F(x, y)$ will be said to *weakly define* a function $f(x)$ (from numbers to numbers) in a system $\mathcal{S}$ if it defines the relation $f(x) = y$, i.e. if for all numbers m and n the following two conditions hold:

(1) If $f(n) = m$, then $F(\overline{n}, \overline{m})$ is provable.
(2) If $f(n) \neq m$, then $F(\overline{n}, \overline{m})$ is refutable.

We will say that $F(x, y)$ *strongly defines* the function $f(x)$ if also, for all m and n:

(3) If $f(n) = m$, the sentence $\forall y(F(\overline{n}, y) \supset y = \overline{m})$ is provable.

Proposition 2. If the formulas of $\Omega 4$ and $\Omega 5$ are provable in $\mathcal{S}$, then any function $f(x)$ weakly definable in $\mathcal{S}$ is strongly definable in $\mathcal{S}$.

Problem 2. Prove Proposition 2.

From Propositions 1 and 2, we have:

Proposition 3. All recursive functions $f(x)$ are strongly definable in any extension $\mathcal{S}$ of (R).

Next we need:

Proposition 4. If $f(x)$ is strongly definable in $\mathcal{S}$, then for any formula $G(x)$, there is a formula $H(x)$ such that for all n, the sentence $H(\overline{n}) \equiv G(f(\overline{n}))$ is provable in $\mathcal{S}$.

Problem 3. Prove Proposition 4.

We say that a function $f(x)$ is $\sum_1$ if the relation $f(x) = y$ is $\sum_1$, and we say that $f(x)$ is *recursive* if the relation $f(x) = y$ is recursive.

Problem 4. Prove that if a function $f(x)$ is $\sum_1$, then it is recursive.

The Diagonal Function

We recall the $\sum_1$ function $r(x, y)$ of Chapter 11, which has the property that for any n which is the Gödel number of a formula $F_n(v_1)$, and any number m, the number $r(n, m)$ is the Gödel number of the sentence $F_n[\overline{m}]$ (which is the sentence $\exists v_1(v_1 = \overline{m} \wedge F_n(v_1))$, which is equivalent to the sentence $F_n(\overline{m})$ (round brackets). We let $d(x)$ be the $\sum_1$ function $r(x, x)$. Thus if n is the Gödel number of $F_n(v_1)$, then $d(n)$ is the Gödel number of the diagonalization $F_n[\overline{n}]$. We call $d(x)$ the *diagonal function*. Since the diagonal function is $\sum_1$, it is recursive (by Problem 4).

Consider now any extension $\mathcal{S}$ of the system (R). Since the relation $d(x) = y$ is recursive, it is definable in $\mathcal{S}$ (by Proposition 1), which means that the function $d(x)$ is weakly definable in $\mathcal{S}$. Since all the formulas of $\Omega 4$ and $\Omega 5$ are provable in $\mathcal{S}$, it then follows by Proposition 2 that $d(x)$ is *strongly* definable in $\mathcal{S}$. We thus have:

Proposition 5. The diagonal function $d(x)$ is strongly definable in every extension $\mathcal{S}$ of (R).

Fixed Points

For the rest of the chapter, for any expression X, we let $\overline{X}$ be *the Peano numeral that designates the Gödel number of* X. Thus for any formula $F(x)$, by $F(\overline{X})$ is meant $F(\overline{n})$, where n is the Gödel number of X.

> *Note:* The notation defined in the previous paragraph is not exactly inconsistent with notation that has been previously used. It's just that we've introduced a new "bar" function, one that operates on expressions, while our old bar function operated on conceptual natural numbers. The context will always make clear which function is intended. Thus the $F(\overline{n})$ here has exactly the same meaning it has always had: If n is a natural number (say 3, or the Gödel number of any expression), then $\overline{n}$ is the Peano numeral corresponding to n (in the case of $n = 2$, $\overline{2}$ would be the Peano numeral $0''$), which is an *expression* in Peano Arithmetic. And $F(\overline{2})$ would be the result of substituting $0''$ for every free occurrence of x in $F(x)$. What may be confusing is that if we took our expression $0''$ as the expression X in the definition above, the statement of the definition would be informing us that $\overline{0''}$ is the Peano numeral that designates the Gödel number of $0''$, i.e. the Peano numeral that designates 12122122 (as expressed in dyadic notation, which is 346 in decimal notation), so that $\overline{0''}$ is actually 0 with 346 strokes following it. But the situation the definition above is most useful for is one we've encountered over and over again, namely one in which X itself is a formula, say $F(x)$, which of course has its Gödel number, which we might denote by f. We often switched to a new notation once we had the Gödel number of a formula: we used $F_f(x)$ for the formula with Gödel number f (which might or might not have x as a free variable occurring within it). What we did frequently with such a formula was to substitute the Peano numeral of the Gödel number of the formula for all free occurrence of the variable x in the formula $F_f(x)$, something we denoted by $F_f(\overline{f})$.

A sentence X is called a *fixed point* of a formula $F(x)$ (with respect to a system $\mathcal{S}$) if the sentence $X \equiv F(\overline{X})$ is provable in $\mathcal{S}$. Thus X is a fixed point of $F(x)$ iff $X \equiv F(\overline{n})$, where n is the Gödel number of X.

Proposition 6. If the diagonal function $d(x)$ is strongly definable in $\mathcal{S}$, then every formula $F(x)$ has a fixed point in $\mathcal{S}$.

Problem 5. Prove Proposition 6.

From Propositions 6 and 5, we have:

Proposition 7. If $\mathcal{S}$ is an extension of (R), then every formula $F(x)$ of $\mathcal{S}$ has a fixed point.

The following is a theorem of Tarski [1933, 1936]:

Theorem T_2. For a consistent system $\mathcal{S}$, if the diagonal function $d(x)$ is strongly definable in $\mathcal{S}$, then the set P_0 of Gödel numbers of the provable sentences of $\mathcal{S}$ is not definable in $\mathcal{S}$.

Problem 6. Prove Theorem T_2.

Truth Predicates

A formula $T(x)$ is called a *truth predicate* for a system $\mathcal{S}$ if for every sentence X, the sentence $X \equiv T(\overline{X})$ is provable in $\mathcal{S}$.

The following is also due to Tarski.

Theorem T_3. If $\mathcal{S}$ is consistent and the diagonal function $d(x)$ is strongly definable in $\mathcal{S}$, then there is no truth predicate for $\mathcal{S}$.

Problem 7. Prove Theorem T_3.

Fixed points play a key role in results related to Gödel's second theorem. They also provide neat alternative ways of proving the Gödel and Rosser incompleteness theorems, as indicated in the following exercise.

Exercise. Consider a system $\mathcal{S}$ in which P_0 is the set of Gödel numbers of the provable sentences of $\mathcal{S}$ and R_0 is the set of Gödel numbers of the refutable sentences of $\mathcal{S}$. Prove the following facts:

(1) If $F(x)$ represents R_0, then any fixed point of $F(x)$ is undecidable (in $\mathcal{S}$, providing $\mathcal{S}$ is consistent).

(2) If $F(x)$ represents P_0 and $\mathcal{S}$ is consistent, then any fixed point of $\sim F(x)$ is undecidable.

(3) If $F(x)$ represents some superset of R_0 disjoint from P_0, then any fixed point of $F(x)$ is undecidable.

(4) If $F(x)$ represents some superset of P_0 disjoint from R_0, then any fixed point of $\sim F(x)$ is undecidable.

(5) Suppose $F(x, y)$ enumerates P_0 in $\mathcal{S}$ and that G is a fixed point of $\forall y \sim F(x, y)$. Then

   (a) If the system is simply consistent, then G is not provable.

   (b) If the system is omega consistent, then G is also not refutable, so is thus undecidable.

Part II. The Unprovability of Consistency

Gödel's second incompleteness theorem, which we will shortly state more concisely, has been generalized and abstracted in many ways and thus has led to the notion of a *provability predicate*, which plays a fundamental role in much modern metamathematical research. To this notion we now turn.

Provability Predicates

A formula $P(x)$ is called a *provability predicate* for a system $\mathcal{S}$ if for all sentences X and Y, the following three conditions hold. [We continue to use the notation $\overline{X}$ for the Peano numeral that designates the Gödel number of X.]

P_1: If X is provable in $\mathcal{S}$, then so is $P(\overline{X})$.
P_2: $P(\overline{X \supset Y}) \supset (P(\overline{X}) \supset P(\overline{Y}))$ is provable in $\mathcal{S}$.
P_3: $P(\overline{X}) \supset P(\overline{P(\overline{X})})$ is provable in $\mathcal{S}$.

Suppose now that $P(x)$ is a $\sum_1$ formula that expresses the set P_0 of the Gödel numbers of the provable sentences of Peano Arithmetic. Under the assumption of omega consistency, $P(x)$ represents the set P_0. Under the weaker assumption of simple consistency, all that follows is that $P(x)$ represents some *superset* of P_0, but that is enough to imply that if X is provable in Peano Arithmetic, then so is $P(\overline{X})$. Therefore property P_1 holds. As for the property P_2, the sentence $P(\overline{X \supset Y}) \supset (P(\overline{X}) \supset P(\overline{Y}))$ is obviously true (since what it says is that if $X \supset Y$ and X are both provable, so is Y, which is of course the case, since modus ponens is an inference rule of Peano Arithmetic. It is not very difficult to formalize this argument and show that the above sentence is not only true, but provable in Peano Arithmetic.

As for property P_3, the sentence $P(\overline{X}) \supset P(\overline{P(\overline{X})})$ asserts that if X is provable, then $P(\overline{X})$ is provable, which is true by property P_1. It is also provable in Peano Arithmetic, but the proof is extremely elaborate and goes beyond the scope of this volume. A sketch of the proof can be found in Chapter 2 of Boolos [1979]. A detailed treatment can be found for a system similar to Peano Arithmetic in Hilbert-Bernays [1934 and 1939].

In what follows, it will be assumed that $P(x)$ is a provability predicate for $\mathcal{S}$, and that all logically valid formulas are provable in $\mathcal{S}$ and that $\mathcal{S}$ is closed under modus ponens.

A provability predicate $P(x)$ for $\mathcal{S}$ satisfies the following conditions (for all sentences X and Y):

P_4: If $X \supset Y$ is provable (in $\mathcal{S}$), so is $P(\overline{X}) \supset P(\overline{Y})$.
P_5: If $X \supset (Y \supset Z)$ is provable, so is $P(\overline{X}) \supset (P(\overline{Y}) \supset P(\overline{Z}))$.
P_6: If $X \supset (P(\overline{X}) \supset Y)$ is provable, so is $P(\overline{X}) \supset P(\overline{Y})$.

Problem 8. Prove P_4, P_5, and P_6.

Properties P_1 and P_6 play the key roles is what follows.

Key Lemma. If X is a fixed point of $P(x) \supset Y$, then $(P(\overline{Y}) \supset Y) \supset X$ is provable in $\mathcal{S}$.

Problem 9. Prove the Key Lemma.

The Unprovability of Consistency

We let f stand for any sentence that is refutable in $\mathcal{S}$ (such as the negation of any tautology, or, for Peano Arithmetic, the sentence $0 = \overline{1}$, which is a common choice). We let f remain fixed for the discussion. We note that since f is refutable, then for any sentence X, the sentence $\sim X \equiv (X \supset f)$ is provable in $\mathcal{S}$. We also note that since f is refutable then f is provable in $\mathcal{S}$ iff $\mathcal{S}$ is inconsistent, and therefore that f is not provable iff $\mathcal{S}$ is consistent.

We let *concis* be the sentence $\sim P(\overline{f})$. If $P(x)$ is a *correct* provability predicate for $\mathcal{S}$, in the sense that for any sentence X, the sentence $P(\overline{X})$ is *true* iff X is provable in $\mathcal{S}$, then the sentence $\sim P(f)$ is true iff f is not provable in $\mathcal{S}$, in other words, iff $\mathcal{S}$ is consistent. However, in what follows, we do not have to assume that $P(x)$ is a *correct* provability predicate for $\mathcal{S}$, but only that it is a provability predicate for $\mathcal{S}$, in other words, only that properties P_1, P_2, and P_3 hold (and hence, so do P_4, P_5, and P_6).

Theorem 1. If G is a fixed point of $\sim P(x)$, then consis $\supset G$ is provable in $\mathcal{S}$.

Problem 10. Prove Theorem 1. [Hint: This is almost immediate from the Key Lemma.]

Theorem 2. If G is a fixed point of $\sim P(x)$ and $\mathcal{S}$ is consistent, then G is not provable in $\mathcal{S}$.

Problem 11. Prove Theorem 2.

We shall call a system *diagonalizable* if every formula $F(x)$ of $\mathcal{S}$ has a fixed point. We have seen that every extension of the system (R) is diagonalizable, and that Peano Arithmetic is an extension of (R), and so Peano Arithmetic is diagonalizable.

Theorem 3 [An Abstract Form of Gödel's Second Incompleteness Theorem]. If $\mathcal{S}$ is diagonalizable, and if $\mathcal{S}$ is consistent, then the sentence *consis* is not provable in $\mathcal{S}$.

Problem 12. Prove Theorem 3.

Discussion

For the case that $\mathcal{S}$ is the system Peano Arithmetic and $P(x)$ is a $\sum_1$ formula expressing the set of Gödel numbers of the provable sentences of Peano Arithmetic, the sentence *consis* is a true sentence (assuming that Peano Arithmetic is consistent), but it is not provable in Peano Arithmetic. This result has been paraphrased, "If arithmetic is consistent, then it cannot prove its own consistency." Unfortunately, there has been a great deal of popular nonsense written about this by writers who obviously do not understand what the matter is all about. We have seen such irresponsible statements as: "By Gödel's second theorem, we can never know whether arithmetic is consistent or not." To see how silly this is, suppose it turned out that the sentence *consis* were provable in Peano Arithmetic, or, to be more realistic, consider a system that can prove its own consistency. Would that be any grounds for trusting the consistency of the system? Of course not! If the system were inconsistent, it could prove every sentence, including *consis*. To trust the consistency of a system on the grounds that it can prove its own consistency would be as foolish as to trust the veracity of a man on the grounds that he claims that he never lies!

The mathematician André Weil once made the humorous remark, "God exists, since arithmetic is consistent. The Devil exists, since we cannot prove it." Delightful as this remark is, it is really not accurate. It's not that we cannot prove that Peano Arithmetic is consistent. It is rather that we cannot prove the consistence of Peano Arithmetic using just the axioms of Peano Arithmetic. In higher order mathematical systems, whose axioms are obviously correct, the sentence *consis* of Peano Arithmetic is provable.

In short, the fact that we have shown that if Peano Arithmetic is consistent then *consis* is not provable in Peano Arithmetic does not constitute the slightest evidence that the consistency of Peano Arithmetic is doubtful!

Henkin Sentences and Löb's Theorem

Leon Henkin [1952] raised the following famous question about the system of Peano Arithmetic. Since the system is diagonalizable, there is a fixed point for $P(x)$ – a sentence H such that $H \equiv P(\overline{H})$, is provable in Peano Arithmetic. Unlike Gödel's sentence G, which is a fixed point of $\sim P(x)$, which is true iff it is not provable, Henkin's sentence H is true iff it *is* provable! This means that it is either both true and provable (in Peano Arithmetic), or false but not provable. Is there any way to tell which? This problem was answered by Martin Hugo Löb [1955], who showed that the provability of even $P(\overline{H}) \supset H$ (so of course $P(\overline{H}) \equiv H$) was enough to guarantee that H was in fact provable in Peano Arithmetic. Here is Löb's Theorem.

Theorem 4 [Löb's Theorem]. For any diagonalizable system $\mathcal{S}$ and provability predicate $P(x)$ for $\mathcal{S}$, and any sentence Y, if $P(\overline{Y}) \supset Y$ is provable in $\mathcal{S}$, so is Y.

Problem 13. Prove Löb's Theorem. [Hint: Use the Key Lemma]

As observed by Georg Kreisel [1965], Gödel's Second Incompleteness Theorem is a special case of Löb's Theorem (the case when Y is the sentence f).

Problem 14. Explain why this is so.

There is so much more to be said about provability predicates! A whole branch of mathematical logic has resulted from tying up provability predicates with the subject known as *modal logic*, which is commonly about *necessary* truth. A pioneer in this direction was George Boolos, who united the two fields in his excellent book *The Unprovability of Consistency: An Essay in Modal Logic* [1979], which we strongly recommend as a follow-up to this chapter.

* * *

This is a good temporary stopping place. The reader has now seen but the beginnings of mathematical logic, which by now has branched out into several areas, such as higher-order logics, recursion theory, model theory, set theory, proof theory, combinatory logic, modal logic, and intuitionistic logic to name a few. There is also much more to First-Order Logic than that in the present volume, which is but the first in a projected series. The next one will concentrate mainly on further topics in First-Order Logic and recursion theory, and possibly some combinatory logic.

Solutions to the Problems of Chapter 14

1. To begin with, let us note the obvious fact that to say that a formula F *defines* a relation R in a system $\mathcal{S}$ is equivalent to saying that F strongly separates R from its complement $\widetilde{R}$. Thus a relation R is definable in $\mathcal{S}$ iff R is strongly separable from $\widetilde{R}$ in $\mathcal{S}$.

   Now suppose $\mathcal{S}$ is an extension of (R). Then $\Omega 4$ and $\Omega 5$ both hold in $\mathcal{S}$. Therefore, as shown in the last chapter (Problem 7), for any formula $F(x)$, and any number n, the formula $(F(\overline{0}) \vee \ldots \vee F(\overline{n})) \supset (\forall x \leq n)F(x)$ is provable in $\mathcal{S}$, and of course, so is $x \leq \overline{n} \vee \overline{n} \leq x$. Therefore, by the separation lemmas of Chapter 10, any two disjoint relations R_1, R_2 (of the same degree) which are enumerable in $\mathcal{S}$, are strongly separable in $\mathcal{S}$. Now, any $\sum_1$ relation R is enumerable in $\mathcal{S}$, since R is the domain of a $\sum_0$ relation S, which is definable in $\mathcal{S}$ because $\mathcal{S}$ is $\sum_0$ complete. If R is a recursive relation, then R and $\widetilde{R}$ are both $\sum_1$, hence both are enumerable in $\mathcal{S}$, hence R is strongly separable from $\widetilde{R}$ in $\mathcal{S}$, which means that R is definable in $\mathcal{S}$.
2. Suppose all formulas of $\Omega 4$ and $\Omega 5$ are provable in $\mathcal{S}$ and that $F(x, y)$ weakly defines $f(x)$ in $\mathcal{S}$. Let $G(x, y)$ be the formula $F(x, y) \wedge \forall z(F(x, z) \supset y \leq z)$.

We will show that $G(x, y)$ *strongly* defines $f(x)$ in $\mathcal{S}$. Suppose $f(n) = m$. We are to show three things:

(1) $G(\overline{n}, \overline{m})$ is provable (in $\mathcal{S}$).

(2) For any $k \neq m$, $G(\overline{n}, \overline{k})$ is refutable (in $\mathcal{S}$).

(3) $\forall y(G(\overline{n}, y) \supset y = \overline{m})$ is provable (in $\mathcal{S}$).

(1) We first show that the formula $z \leq \overline{m} \supset (F(\overline{n}, z) \supset \overline{m} \leq z)$ is provable. Well, suppose $F(\overline{n}, \overline{m})$ is provable, and further assume that $k \leq m$. Then either $k < m$ or $k = m$. If the former, then $k \neq m$, hence $F(\overline{n}, \overline{k})$ is refutable. If the latter, then $\overline{m} \leq \overline{k}$ is provable (since $\overline{m} \leq \overline{m}$ by $\Omega 5$ and Proposition 1(b), both of Chapter 13). Thus either $F(\overline{n}, \overline{k})$ is refutable or $\overline{m} \leq \overline{k}$ is provable, and in either case $F(\overline{n}, \overline{k}) \supset \overline{m} \leq \overline{k}$ is provable.

Since $F(\overline{n}, \overline{k}) \supset \overline{m} \leq \overline{k}$ is provable for all $k \leq n$, then, by $\Omega 4$, the following is provable:

(a) $z \leq \overline{m} \supset (F(\overline{n}, z) \supset \overline{m} \leq z)$

Also, of course, the following tautology is provable:

(b) $\overline{m} \leq z \supset (F(\overline{n}, z) \supset \overline{m} \leq z)$

By (a), (b) and $\Omega 5$, the formula $F(\overline{n}, z) \supset \overline{m} \leq z$ is provable, hence the sentence $\forall z(F(\overline{n}, z) \supset \overline{m} \leq z)$ is provable (by Proposition 1(a) of Chapter 13). Also, we are assuming that $F(\overline{n}, \overline{m})$ is provable, and as a consequence we see that the sentence $F(\overline{n}, \overline{m}) \wedge \forall z(F(\overline{n}, z) \supset \overline{m} \leq z)$ – which is the sentence $G(\overline{n}, \overline{m})$ – is provable in $\mathcal{S}$. This proves (1).

(2) This is pretty obvious: If $k \neq m$, then $F(\overline{n}, \overline{k})$ is refutable, hence so is $F(\overline{n}, \overline{k}) \wedge \forall z(F(\overline{n}, z) \supset \overline{k} \leq z)$, which is the sentence $G(\overline{n}, \overline{k})$.

(3) The formula $G(\overline{n}, y) \supset \forall z(F(\overline{n}, z) \supset y \leq z)$ is a tautology, hence provable is $\mathcal{S}$. Also, $\forall z(F(\overline{n}, z) \supset y \leq z) \supset (F(\overline{n}, \overline{m}) \supset y \leq \overline{m})$ is provable (it is an axiom of First-Order Arithmetic). Thus, by Propositional Logic, the following is provable.

(a) $G(\overline{n}, y) \supset y \leq \overline{m}$

Next, we note that for any $k \leq m$ the sentence $G(\overline{n}, \overline{k}) \supset \overline{k} = \overline{m}$ is provable, because if $k < m$, then $G(\overline{n}, \overline{k})$ is refutable (since $F(\overline{n}, \overline{k})$ is) and if $k = m$ then $\overline{k} = \overline{m}$ is provable. Then by $\Omega 4$ the following is provable:

(b) $y \leq \overline{m} \supset (G(\overline{n}, y) \supset y = \overline{m})$

From (a) and (b), by Propositional Logic, the formula $G(\overline{n}, y) \supset y = \overline{m}$ is provable. Thus so is $\forall y(G(\overline{n}, y) \supset y = \overline{m})$ (by Proposition 1(a), Chapter 13).

This concludes the proof.

3. Suppose $F(x, y)$ strongly defines the function $f(x)$. Given a formula $G(x)$ we take $H(x)$ to be the formula $\exists y(F(x, y) \wedge G(y))$. Now, suppose that $f(n) = m$.

We are to show that $H(\overline{n}) \equiv G(\overline{m})$ is provable. Thus we are to show that $G(\overline{m}) \supset H(\overline{n})$ and $H(\overline{n}) \supset G(\overline{m})$ are both provable

(1) To show that $G(\overline{m}) \supset H(\overline{n})$ is provable, we first note that since $F(x, y)$ defines $f(x)$ and $f(n) = \text{m}$, then $F(\overline{n}, \overline{m})$ is provable. Then by Propositional Logic, $G(\overline{m}) \supset (F(\overline{n}, \overline{m}) \wedge G(\overline{m}))$ is provable. Hence, by First-Order Logic, $G(\overline{m}) \supset \exists y(F(\overline{n}, y) \wedge G(y))$, which is $G(\overline{m}) \supset H(\overline{n})$, is provable.

(2) In the other direction, since $F(x, y)$ *strongly* defines $f(x)$, then

  (a) $\forall y(F(\overline{n}, y) \supset y = \overline{m})$ is provable, hence so is the open formula
  (b) $F(\overline{n}, y) \supset y = \overline{m}$. Then, by Propositional Logic, the formula
  (c) $(F(\overline{n}, y) \wedge G(y)) \supset (y = \overline{m} \wedge G(y))$ is provable. Also,
  (d) $(y = \overline{m} \wedge G(y)) \supset G(\overline{m})$ is provable (by First-Order Arithmetic), and so, by modus ponens (Propositional Logic),
  (e) $(F(\overline{n}, y) \wedge G(y)) \supset G(\overline{m})$ is provable. Then, by First-Order Logic,
  (f) $\exists y(F(\overline{n}, y) \wedge G(y)) \supset G(\overline{m})$ is provable. And that formula is
  (g) $H(\overline{n}) \supset G(\overline{m})$.

4. Suppose $f(x)$ is $\sum_1$. Thus the relation $f(x) = y$ is $\sum_1$. We are to show that the relation $f(x) \neq y$ is also $\sum_1$. Well, $f(x) \neq y$ iff $\exists z(f(x) = z \wedge z \neq y)$. The condition $f(x) = z \wedge z \neq y$ is $\sum_1$, so the relation $\exists z(f(x) = z \wedge z \neq y)$ is $\sum_1$ (by Problem 10 of Chapter 11).

5. Let $F(x)$ be a formula that strongly defines $d(x)$ in $\mathcal{S}$. By Proposition 4, there is a formula $H(v_1)$ such that for all numbers n, the sentence $H(\overline{n}) \equiv F(d(\overline{n}))$ is provable in $\mathcal{S}$, hence so is the sentence $H[\overline{n}] \equiv F(d(\overline{n}))$. Let h be the Gödel number of $H(v_1)$ Then $H[\overline{h}] \equiv F(d(\overline{h}))$ is provable in $\mathcal{S}$, and so $H[\overline{h}]$ is a fixed point of $F(x)$, since $d(h)$ is the Gödel number of $H[\overline{h}]$.

6. Some preliminaries first: Suppose there is a formula $F(x)$ that defines P_0 in $\mathcal{S}$. Thus for all numbers n:

(1) If $n \in P_0$, then $F(\overline{n})$ is provable.
(2) If $n \notin P_0$, then $F(\overline{n})$ is refutable.

For any Gödel number n of a sentence S_n, the number n is in P_0 iff S_n is provable. Thus:

(1′) If S_n is provable, then $F(\overline{n})$ is provable.
(2′) If S_n is not provable, then $F(\overline{n})$ is refutable.

Now suppose that the diagonal function $d(x)$ is strongly definable in $\mathcal{S}$. Then there is a fixed point S_n for $\sim F(x)$. Thus $S_n \equiv \sim F(\overline{n})$ is provable. Hence S_n is provable iff $F(\overline{n})$ is refutable, and so by (1′), (2′):

(1″) If $F(\overline{n})$ is refutable, then $F(\overline{n})$ is provable.
(2″) If $F(\overline{n})$ is provable, then $F(\overline{n})$ is refutable.

From (2″) it follows that $F(\overline{n})$ is refutable, and then by (1″), that $F(\overline{n})$ is provable, which means that the system $\mathcal{S}$ is inconsistent. Therefore if $\mathcal{S}$ is consistent, it cannot be that P_0 is definable in $\mathcal{S}$ and $d(x)$ is strongly definable in $\mathcal{S}$.

7. Suppose $\mathcal{S}$ is consistent, and that the diagonal function $d(x)$ is strongly definable in $\mathcal{S}$. Now consider any formula $F(x)$. There must be a fixed point X of $\sim F(x)$. Then X cannot also be a fixed point of $F(x)$ (for then, $F(\overline{X}) \equiv \sim F(\overline{X})$ would be provable, contrary to the assumption that $\mathcal{S}$ is consistent). Therefore it is not the case that $X \equiv F(\overline{X})$ is provable in $\mathcal{S}$, and hence $F(x)$ cannot be a truth predicate.
8. (1) To prove P_4: Suppose $X \supset Y$ is provable. Then so is $P(\overline{X \supset Y})$ by Property P_1. Hence so is $P(\overline{X}) \supset P(\overline{Y})$ by Property P_2 and modus ponens.
   (2) To prove P_5: Suppose $X \supset (Y \supset Z)$ is provable. Then by P_4 we have
   (a) $P(\overline{X}) \supset P(\overline{Y \supset Z})$ is provable. Also, by P_2:
   (b) $P(\overline{Y \supset Z}) \supset (P(\overline{Y}) \supset P(\overline{Z}))$ is provable.
   That $P(\overline{X}) \supset (P(\overline{Y}) \supset P(\overline{Z}))$ is provable follows from (a) and (b) by Propositional Logic.
   (3) To prove P_6: Suppose that $X \supset (P(\overline{X}) \supset Y)$ is provable. Then by P_5, the following is provable:
   (a) $P(\overline{X}) \supset (P(\overline{P(\overline{X})}) \supset P(\overline{Y}))$.
   Since also $P(\overline{X}) \supset P(\overline{P(\overline{X})})$ is provable by P_3, then by Propositional Logic it follows that $P(\overline{X}) \supset P(\overline{Y})$ is provable.
9. Suppose X is a fixed point of $P(x) \supset Y$. Thus:
   (1) $X \equiv (P(\overline{X}) \supset Y)$ is provable. Then the following are successively provable.
   (2) $X \supset (P(\overline{X}) \supset Y)$ (by (1)).
   (3) $P(\overline{X}) \supset P(\overline{Y})$ (by (2) and Property P_6).
   (4) $(P(\overline{Y}) \supset Y) \supset (P(\overline{X}) \supset Y)$ (from (3) by Propositional Logic).
   (5) $(P(\overline{Y}) \supset Y) \supset X$ (from (1) and (4) by Propositional Logic).
10. Suppose G is a fixed point of $\sim P(x)$. Since $\sim P(x) \equiv (P(x) \supset f)$, then G is a fixed point of $P(x) \supset f$. Thus $G \equiv (P(\overline{G}) \supset f)$ is provable. Then by the Key Lemma, taking G for X and f for Y, the sentence $(P(\overline{f}) \supset f) \supset G$ is provable. But $(P(\overline{f}) \supset f) \equiv \sim P(\overline{f})$ is also provable, and so $\sim P(\overline{f}) \supset G$ is provable. Thus *consis* $\supset G$ is provable.
11. Suppose G is a fixed point of $\sim P(x)$. Thus $G \equiv \sim P(\overline{G})$ is provable. Then if G were provable, the sentence $\sim P(\overline{G})$ would be provable, and also $P(\overline{G})$ (by P_1). Hence $\mathcal{S}$ would be inconsistent. Therefore if $\mathcal{S}$ is consistent, then G is not provable.
12. Suppose $\mathcal{S}$ is diagonalizable. Then there is a fixed point G for $\sim P(x)$. Hence by Theorem 1, the sentence *consis* $\supset G$ is provable in $\mathcal{S}$. Therefore, if *consis* were provable, G would be provable, and $\mathcal{S}$ would then be inconsistent by Theorem 2 (since G is a fixed point of $\sim P(x)$). Thus if $\mathcal{S}$ is consistent, then *consis* is not provable in $\mathcal{S}$.
13. Suppose $P(\overline{Y}) \supset Y$ is provable. Let X be a fixed point of $P(x) \supset Y$. Thus the following are successively provable:

(1) $P(\overline{Y}) \supset Y$ (Given.)

(2) $X \equiv (P(\overline{X}) \supset Y)$ (X is a fixed point of $P(x) \supset Y$). Then the following are successively provable:

(3) $(P(\overline{Y}) \supset Y) \supset X$ (from (2), by the Key Lemma).

(4) X (by (1) and (3)).

(5) $P(\overline{X})$ (by (4) and Property P_1 of provability predicates).

(6) $P(\overline{X}) \supset Y$ (by (4) and (2)).

(7) Y (by (5) and (6)).

14. By Löb's Theorem, taking f for Y, if $P(\overline{f}) \supset f$ is provable, so is f. Thus if *consis* is provable, so is f (since *consis* $\equiv P(\overline{f}) \supset f$ is provable). But if f is provable, the system is inconsistent. Hence if the system is consistent, then the sentence *consis* is not provable in the system.

References

Boole, George, *An Investigation of the Laws of Thought*, reprinted by Cambridge University Press, 2009.

Boolos, George, *The Unprovability of Consistency: An Essay in Modal Logic*, Cambridge University Press, 1979.

Brouwer, L. E. J., "On the Domains of Definition of Functions" [1927]. published in van Heijenoort, Jean, ed., *From Frege to Gödel: A Source Book in Mathematical Logic, 1879–1931*, Harvard Univ. Press, pp. 199–215. 1967.

Cantor, Georg, "Grundlagen einer allgemeinen Mannigfaltigkeitslehre" ("Foundations of a General Theory of Aggregates"), *Mathematische Annalen*, 1883.

Church, Alonzo, *Introduction to Mathematical Logic*, Princeton University Press, 1956.

Fraenkel, A. A., *Abstract Set Theory*, 2nd Edition, North Holland, 1961.

Frege, Gottlob, *Begriffsschrift, eine der arithmetischen nachgebildete Formelsprache des reinen Denkens*. Halle a. S.: Louis Nebert, 1879. Translation: *Concept Script, a formal language of pure thought modelled upon that of arithmetic*, by S. Bauer-Mengelberg in Jean Van Heijenoort, ed., 1967. *From Frege to Gödel: A Source Book in Mathematical Logic, 1879–1931*. Harvard University Press.

Gödel, Kurt, Über formal unentsheidbare Sätze der "Principia Mathematica" und verwandter Systeme I, *Monatshefte für Mathematik und Physik*, 1931. Vol. 38. pp. 173–198.

Henken, Leon, "A problem concerning provability, problem 3," *Journal of Symbolic Logic*, 1952. Vol. 17, p. 160.

Hilbert, David and Bernays, Paul, *Foundations of Mathematics* (*Grundlagen der Mathematik*), Springer Verlag, 1939. Vol. 1 1934, Vol. 2.

Jerislow, R. G., "Redundancies in the Hilbert-Bernays derivability conditions for Gödel's second incompleteness theorem," *Journal of Symbolic Logic*, 1973. Vol. 38, pp. 358–367.

Kleene, Stephen Cole, *Introduction to Metamathematics*, D. Van Nostrand Company, Inc. 1952.

Kreisel, Georg and Sacks, Gerald E., "Metarecursive Sets," *Journal of Symbolic Logic*, 1965. Vol. 30 (3):318–338.

Leblanc, Hugues and Snyder, D. Paul, "Duals of Smullyan Trees," *Notre Dame Journal of Formal Logic*, 1972. Vol. 13(3):387–393.

Lukasiewicz, Jan, *Selected Writings*. North-Holland, Edited by L. Borowski. 1970.

Löb, Martin Hugo, "Solution of a problem of Leon Henkin," *Journal of Symbolic Logic*, 1955. Vol. 20, Number 1, pp. 115–18.

Mendelson, Elliott, *Introduction to Mathematical Logic*, Wadsworth and Brooks, 1987.

Peano, Giuseppe, "Sul concetto di numero," *Rivista di Matematica*, 1891. *Vol.* 1, pp. 87–102.

Quine, W. V., "Concatenation as a Basis for Arithmetic," *The Journal of Symbolic Logic*, 1946. Vol. 11, #4, pp. 105–114.

Robinson, Raphael M., "An essentially undecidable axiom system," *Proceedings of the International Congress of Mathematicians*, Cambridge University Press, 1950, pp. 729–730.

Rosser, J. Barkley, "Extensions of some Theorems of Gödel and Church," *Journal of Symbolic Logic*, 1936. Vol. 1, pp. 87–91.

Rosser, J. Barkley, *Logic for Mathematicians*, McGraw Hill, 1953.

Smullyan, Raymond, *Theory of Formal Systems*, Princeton University Press. 1961.

Smullyan, Raymond, *First-Order Logic*, Springer Verlag, 1978.

Smullyan, Raymond, *Gödel's Incompleteness Theorems*, Oxford University Press. 1992.

Smullyan, Raymond, *The Magic Garden of George B. And Other Logic Puzzles*, Polimetrica International Scientific Publisher Monza, Italy, 2007.

Smullyan, Raymond, *Logical Labyrinths*, A. K. Peters, Ltd., 2009.

Smullyan, Raymond, *The Gödelian Puzzle Book*, Dover, 2013.

Tarski, Alfred, "Pojęcie prawdy w językach nauk dedukcyjnych", Towarszystwo Naukowe Warszawskie, Warszawa, 1933. (Text in Polish in the Digital Library WFISUW-IFISPAN-PTF.) The 1936 German translation was titled "Der Wahrheitsbegriff in den formalisierten Sprachen", ("the concept of truth in formalized languages"), sometimes shortened to "Wahrheitsbegriff". An English translation had to await the 1956 first edition of the volume *Logic, Semantics, Metamathematics.*

Tarski, Alfred, "Der Wahrsheitsbegriff in den formalisierten Sprachen der deductiven Disziplinen," *Studia Philosophica*, 1936. *Vol.* 1, pp. 261–405.

Tarski, Alfred, *Undecidable Theories*, North Holland Publishing Company, 1953.

Whitehead, Alfred North and Russell, Bertrand, *Principia Mathematica*, Cambridge University Press, 1910. Vol. 1.

Zermelo, Ernst (1908), "Untersuchungen über die Grundlagen der Mengenlehre I", Mathematische Annalen 65 (2): 261–281. English translation: Heijenoort, Jean van, "Investigations in the foundations of set theory", From Frege to Gödel: A Source Book in Mathematical Logic, 1879–1931, Harvard Univ. Press, pp. 199–215. 1967.

Zwicker, William S., "Playing Games with Games: The Hypergame Paradox," The American Mathematical Monthly, Vol. 94, No. 6, Jun.–Jul., 1987, pp. 507–514.

Index

INDEX

INDEX

INDEX

INDEX

INDEX